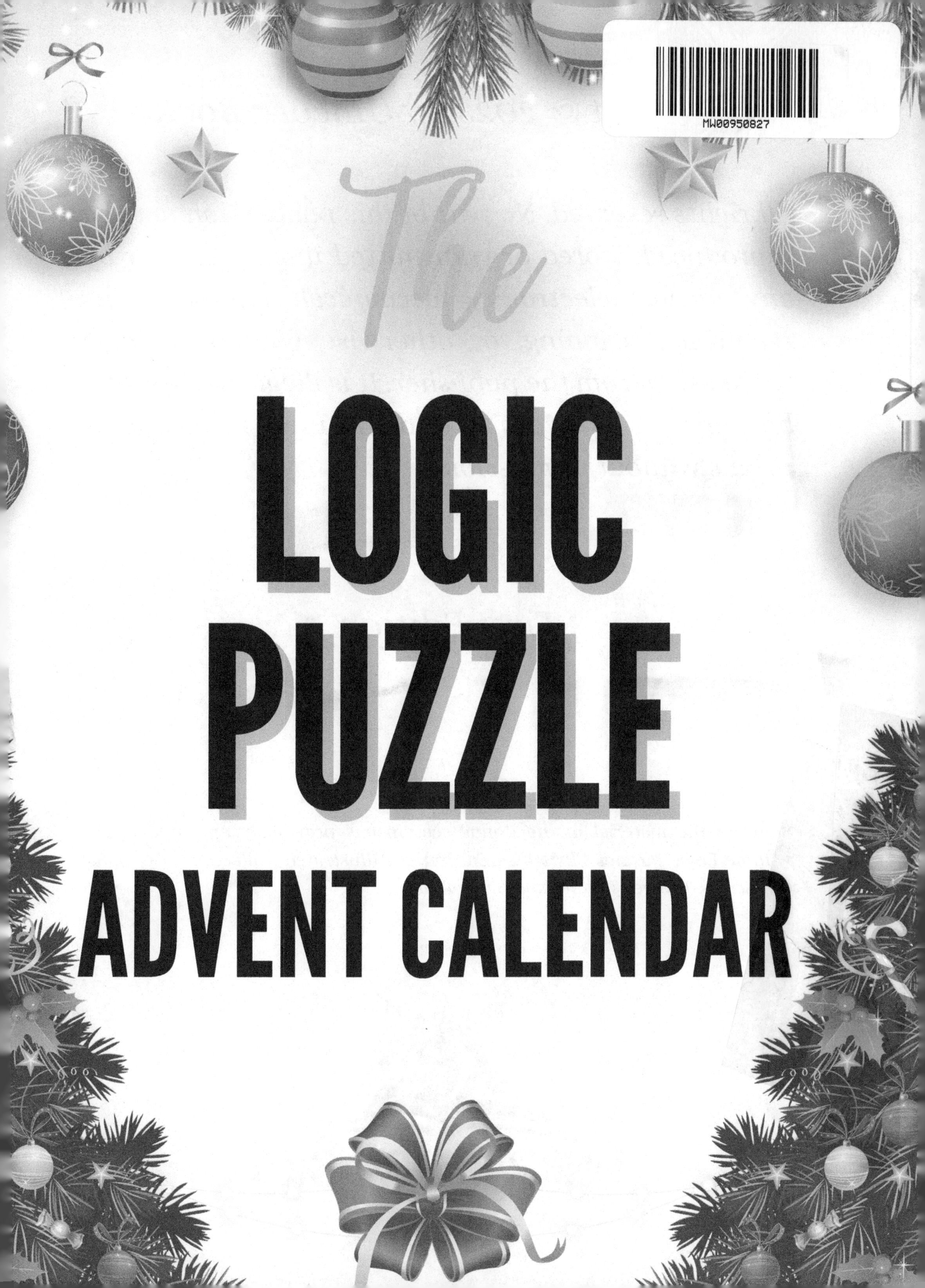

The LOGIC PUZZLE ADVENT CALENDAR

First Edition

*Some of the material in the Bonus section has previously appeared in the volume **Logic Puzzles Christmas Edition**: An Illustrated Collection of Original Christmas-Themed Riddles and Brain Teasers.*

INTRODUCTION

Welcome to this collection of **medium to extremely challenging puzzles** designed to test and sharpen your logical prowess.

From December 1 to 25, you're invited to dive into a daily adventure, exploring a myriad of logical challenges. From **intricate logic grids** of various sizes to **old-school gridless puzzles**, this book will keep your brain engaged and entertained throughout the holiday season.

Each puzzle in this book is **wrapped in the spirit of Christmas**, bringing to life a wondrous snowy world adorned with twinkling lights, Santa's presents, candy canes, cheerful elves, and majestic reindeer.

So, grab your favorite mug, fill it with steaming hot chocolate, and prepare to immerse yourself in the magical atmosphere of the season.

For those seeking additional guidance, we've got you covered. On page 54, you'll discover a QR code that grants you access to a comprehensive bonus guide. This guide not only offers a refresher on the basics of Logic Grid Puzzles but also provides advanced solving techniques to help you conquer even the most challenging puzzles with ease.

In the spirit of Christmas, a season of sharing and joy, we've added an exclusive **bonus section** filled with brain teasers and puzzles that are perfect for sharing with your loved ones, even the littlest ones in the family. Turn to page 61 to discover these special Christmas-themed puzzles meant to be enjoyed together with your family members.

ACKNOWLEDGEMENTS

The Editor extends sincere gratitude to Bethany Nuckolls for testing several puzzles, sharing her invaluable feedback, and contributing four fun-filled puzzles to this book.

THE GREAT CHRISTMAS CHALLENGE

Here is what awaits you on Christmas day:

A colossal logic grid with **735 entries**. Will you manage to crack it?

		ACTIVITY	YEAR	MAIN COURSE	PHOTO ALBUMS	BOARD GAME
		Bake Christmas cookies, Create Christmas crafts, Decorate the tree, Play in the snow, Read Christmas stories, Scavenger hunt, Write letters to Santa		Baked ham, Beef Wellington, Lobster thermidor, Prime rib roast, Roast lamb, Seafood paella, Stuffed pork loin		Dixit, Monopoly, Pictionary, Risk, Scrabble, Ticket to Ride, Uno
RELATIVE	Aunt Mary, Aunt Patricia, Cousin Linda, Grandma Evelyn, Grandpa Victor, Uncle Max, Uncle Richard					
BOARD GAME	Dixit, Monopoly, Pictionary, Risk, Scrabble, Ticket to Ride, Uno					
PHOTO ALBUMS						
MAIN COURSE	Baked ham, Beef Wellington, Lobster thermidor, Prime rib roast, Roast lamb, Seafood paella, Stuffed pork loin					
YEAR						

All the puzzles have been carefully tested and proofread.

Every puzzle has a **unique solution**, which can be obtained **logically** without guesswork from the given clues.

If you are stuck, email us at

unicornbookspuzzles@gmail.com

and we will be happy to provide a hint.

DECEMBER 1

THE CHRISTMAS PRANK

While Santa was resting, three elves stealthily removed all the labels from a selection of presents replacing them with numbers from 1 to 9.

As Santa discovered the prank, he let out a hearty laugh, appreciating the creativity of his little helpers.

Help Santa match each present with its rightful recipient using the clues left behind by the clever elves. Left and right are from your point of view.

Clues

1) Mary's present has an even number, but it is not next to Kyle's present.
2) There are exactly two boxes between Lia's present and Ruth's present.
3) There are exactly two boxes between Ginny's present and Lia's present.
4) Chris's present is somewhere to the left of Lia's present.
5) Zoe's present is somewhere to the left of Ginny's present, which has a number half that of John's.

CHILDREN: Bob, Chris, Ginny, John, Kyle, Lia, Mary, Ruth, Zoe.

_____ _____ _____ _____ _____ _____ _____ _____ _____

If you don't know how to start, there is a hint on the next page.

To check your answers and celebrate your victories, go to page 55.

THE NORTH POLE SLEIGH RACE

Every year, the magical spirit of Christmas comes alive at the North Pole with an event that brings joy and excitement to all: the Great North Pole Sleigh Race.

Each reindeer is paired with an elf partner responsible for decorating the sleigh and riding alongside the reindeer. The elves take this responsibility seriously, adorning each sleigh with festive decorations that reflect the Christmas atmosphere.

Solve the puzzle to discover the placement of each reindeer-elf team in this year's race, the color of their sleighs, and how they were decorated.

Clues

1) Dancer and his golden sleigh crossed the finish line just ahead of Twinkle and his reindeer.
2) Rudolph and Sparkle, both on the same team, finished the race right behind Blitzen.
3) Comet, with a holly-decorated sleigh, crossed the finish line just after Pudding and right before the silver sleigh.
4) Merry adorned his sleigh with colorful ribbons.
5) The sleigh that won the race was not decorated with ornaments.
6) Jingles, riding a red sleigh, finished the race right before the blue sleigh, adorned with bells, but right after Vixen.

THE CHRISTMAS PRANK – HINT

Start by figuring out which present is Ginny's.

		Jingles	Merry	Pudding	Sparkle	Twinkle	Blue	Gold	Red	Silver	White	1	2	3	4	5	Bells	Holly	Lights	Ornaments	Ribbons
REINDEER	Blitzen																				
	Comet																				
	Dancer																				
	Rudolph																				
	Vixen																				
DECORATION	Bells																				
	Holly																				
	Lights																				
	Ornaments																				
	Ribbons																				
PLACEMENT	1																				
	2																				
	3																				
	4																				
	5																				
SLEIGH COLOR	Blue																				
	Gold																				
	Red																				
	Silver																				
	White																				

Column groups: ELF (Jingles, Merry, Pudding, Sparkle, Twinkle); SLEIGH COLOR (Blue, Gold, Red, Silver, White); PLACEMENT (1, 2, 3, 4, 5); DECORATION (Bells, Holly, Lights, Ornaments, Ribbons)

REINDEER	ELF	SLEIGH COLOR	PLACEMENT	DECORATION
Blitzen				
Comet				
Dancer				
Rudolph				
Vixen				

DECEMBER 2

HO-HO-HO AND WHO LIVES BELOW

Santa Claus stood on the roof, his brow furrowed in confusion. 'These houses with numerous chimneys are always so annoying,' he muttered. Determined not to make any mistakes, Santa consulted his notes. Each chimney represents a different home, numbered 1, 2, and 3 on Chuckleberry Drive. Help Santa match each family to their respective chimney. Moreover, determine the name and age of the child in each family and the present they are expecting from Santa. (If you don't know how to start, there is a hint on the next page.)

Clues

1) The child of the Lees, who will receive a book, is not called Ben.
2) The two chimneys Santa will use to deliver Ben's present and the baseball bat are next to each other.
3) Chimney 1 does not belong to the Kings.
4) The child who will receive a baseball bat does not live at number 2.
5) The 8-year-old child does not live in a house with a number higher by one than the child who will receive a board game.
6) The Wards have a 7-year-old child. Their chimney is next to the Lees'.
7) Roy is six years old.

FAMILIES: Lee, King, Ward.

CHILDREN: Adam, Ben, Roy.

AGES: 6, 7, 8.

PRESENTS: Board game, Book, Baseball bat.

	1	2	3
FAMILY	______	______	______
CHILD	______	______	______
AGE	______	______	______
PRESENT	______	______	______

CAROLS AND CHARITY

A group of five friends went from house to house singing Christmas carols and collecting donations for charity. Five generous families not only made a monetary contribution but also treated the children to homemade Christmas sweets. Determine the specific carol sung to each family, the amount donated by each family, and the type of sweets offered to the children.

Clues

1) The family that enjoyed a performance of "O Holy Night" donated more than the Millers but less than the family who offered the children freshly baked cookies.

2) Either the Smiths or the family who gave the children cookies listened to "Jingle Bells." The other family donated $10 less than the one who enjoyed "Jingle Bells."

3) The Wilsons donated more than the Browns.

4) The Garcias donated $10 less than the family that offered the kids toffees.

5) The family that gave the children pecan pie bars donated more than the family that gave them eggnog fudge.

6) The Browns enjoyed a performance of their favorite Christmas song: "Feliz Navidad."

7) The family that donated $60 did not give the children pecan pie bars.

8) The family that gave the children eggnog fudge, not the Smiths, listened to "Silent Night."

9) The Browns, the family that donated $30, and the family that gave the kids rum balls are three different families.

10) The family who listened to "Joy to the World," not the Millers, donated $40.

		SONG					SWEETS					DONATION ($)				
		Feliz Navidad	Jingle Bells	Joy to the World	O Holy Night	Silent Night	Cookies	Eggnog fudge	Pecan pie bars	Rum balls	Toffees	30	40	50	60	70
FAMILY	Brown															
	Garcia															
	Miller															
	Smith															
	Wilson															
DONATION ($)	30															
	40															
	50															
	60															
	70															
SWEETS	Cookies															
	Eggnog fudge															
	Pecan pie bars															
	Rum balls															
	Toffees															

FAMILY	SONG	SWEETS	DONATION ($)
Brown			
Garcia			
Miller			
Smith			
Wilson			

HO-HO-HO AND WHO LIVES BELOW - HINT

Start by figuring out which of the three is the chimney of the King family.

DECEMBER 3

THE MERRY HOSTS

During the Christmas holidays, a close-knit group of seven teenage friends decided to make the most of their time off. Throughout last week, from Monday to Sunday, they gathered at one friend's house each day. The host of the day warmly welcomed the group, preparing delightful snacks and choosing their favorite Christmas movie for a cozy movie night.

Solve the puzzle to discover the identity of the host for each day, the snacks they prepared for their friends, and the Christmas movies they shared.

Clues

1) The friends watched "Elf" the day before going to Trevor's house.
2) The friends watched "Home Alone" the day after watching "Jingle All the Way."
3) At Rachel's house, the friends didn't watch "Jingle All the Way."
4) Katherine offered her friends either cinnamon pretzels or gingerbread cookies.
5) The friends watched "Rise of the Guardians" earlier in the week than the day they ate cranberry brie bites.
6) Alice hosted earlier in the week than the day they watched "The Grinch."
7) The seven friends are Jack, Nora, Alice, the friend who prepared fruit skewers, the one who hosted on Wednesday, and the two friends who suggested watching "Home Alone" and "Jingle All the Way."
8) The friend who made peppermint popcorn hosted the group three days before the friend who baked gingerbread cookies.
9) On Tuesday, the friends watched "The Polar Express."
10) The friends ate mini meatballs five days after watching "Arthur Christmas."
11) Jack hosted earlier in the week than the friend who prepared stuffed mushrooms.
12) The person who baked stuffed mushrooms is either Alice or the friend whose favorite movie is "Rise of the Guardians."
13) On Friday, the friends either watched "Elf" or ate fruit skewers, but not both.
14) Of Katherine and the friend who offered cranberry brie bites, one hosted on Sunday, and the other suggested watching "The Grinch."

		CHRISTMAS MOVIE							DAY							SNACK						
		Arthur Christmas	Elf	Home Alone	Jingle All the Way	Rise of the Guardians	The Grinch	The Polar Express	Monday	Tuesday	Wednesday	Thursday	Friday	Saturday	Sunday	Cinnamon pretzels	Cranberry brie bites	Fruit skewers	Gingerbread cookies	Mini meatballs	Peppermint popcorn	Stuffed mushrooms
HOST	Alice																					
	Jack																					
	Katherine																					
	Nora																					
	Rachel																					
	Sam																					
	Trevor																					
SNACK	Cinnamon pretzels																					
	Cranberry brie bites																					
	Fruit skewers																					
	Gingerbread cookies																					
	Mini meatballs																					
	Peppermint popcorn																					
	Stuffed mushrooms																					
DAY	Monday																					
	Tuesday																					
	Wednesday																					
	Thursday																					
	Friday																					
	Saturday																					
	Sunday																					

HOST	CHRISTMAS MOVIE	DAY	SNACK
Alice			
Jack			
Katherine			
Nora			
Rachel			
Sam			
Trevor			

DECEMBER 4

ELVES UNSCRAMBLED

December is definitely the busiest month of the year for Santa and his little helpers. That is why they have regular staff meetings to ensure everything is proceeding smoothly. While the elves might all look very similar, they have very different tasks and responsibilities. Each of the six elves in front of Santa is responsible for a different continent; moreover, each has a different role in the gift laboratory. Use the clues to find out the name of each elf, the continent for which they are responsible, and their role. Left and right are from your point of view.

Clues

1) Tinsel, who is responsible for Africa, is denoted by a number twice that of Frosty.
2) Elf number 3 is responsible for Oceania.
3) Jolly, tasked with sorting and labeling the gifts, is immediately right of the elf responsible for Europe.
4) Elf number 5 is Merry. He is not tasked with quality control.
5) The elf who performs quality control is immediately to the left of the elf who loads the sleigh.
6) Sparkle is standing immediately to the left or immediately to the right of the elf who assembles the toys, who is denoted by an odd number.
7) Of the two elves tasked with loading the sleigh and wrapping the gifts, one is responsible for North America. Of these two elves, the one responsible for North America is somewhere to the right of Twinkle, while the other is somewhere to the left of Twinkle.
8) The elf tasked with decorating is responsible for Asia.

NAMES: Jolly, Merry, Sparkle, Frosty, Twinkle, Tinsel.

CONTINENTS: Asia, Africa, Europe, North America, South America, Oceania

ROLES: Decorating, Gift wrapping, Quality control, Sleigh loading, Sorting and labeling, Toy assembly.

	1	2	3	4	5	6
NAME	______	______	______	______	______	______
CONTINENT	______	______	______	______	______	______
ROLE	______	______	______	______	______	______

DECORATIVE DILEMMAS

It's time to decorate the Christmas tree! Five siblings have decided to find out who can place more ornaments on the tree in 10 minutes. Solve the puzzle to discover the type and color of the ornaments each sibling placed on the tree and how many they managed to set within the given ten minutes of time.

Clues

1) Alex placed 10 fewer ornaments than Bella but five fewer ornaments than the child who placed the green ornaments.
2) There are 10 fewer figurines on the tree than silver ornaments.
3) The five children are Bella, Chad, Luis, the child who placed the icicles, and the child who placed the blue ornaments.
4) Luis placed either the red ornaments or the egg-shaped ornaments on the tree.
5) Of Chad and the child who placed 34 ornaments, one dealt with figurines and the other with red ornaments.
6) Bella's ornaments were not yellow.
7) The child who placed 39 ornaments is not Luis.
8) Alex placed either the yellow ornaments or the globes on the tree.
9) Naomi placed the icicles on the tree.

		ORNAMENTS					COLOR					NUMBER				
		Angels	Eggs	Figurines	Globes	Icicles	Blue	Green	Red	Silver	Yellow	24	29	34	39	44
SIBLING	Alex															
	Bella															
	Chad															
	Luis															
	Naomi															
NUMBER	24															
	29															
	34															
	39															
	44															
COLOR	Blue															
	Green															
	Red															
	Silver															
	Yellow															

SIBLING	ORNAMENTS	COLOR	NUMBER
Alex			
Bella			
Chad			
Luis			
Naomi			

ELVES UNSCRAMBLED – HINT

Start by figuring out the name of elf number 1.

DECEMBER 5

HOLIDAY HOTCHOCO HEIST

In December, at the coffee shop Chocolica, the aroma of freshly brewed coffee mingles with the festive scent of holiday spices.

To celebrate the Christmas season, Chocolica has introduced six decadent beverages with names like Mistletoe Mirage and Snowfall Sip. These concoctions are not your ordinary hot chocolates; they are crafted using various types of chocolates and a splash of different liquors, creating a unique blend of flavors designed to warm your heart and lift your spirits.

Determine the ingredients and pricing of each beverage using the clues provided below. Left and right are from your point of view.

Clues

1) The Frosty Elixir is made with dark chocolate.
2) The beverage with Baileys Irish Cream is more than $2 cheaper than the Frosty Elixir.
3) The beverage with Amaretto costs twice as much as the Mistletoe Mirage.
4) The beverage with white chocolate costs $2 less than the one with rum.
5) The beverage with orange chocolate is not exactly $2 more expensive than the Winter Dream.
6) The spicy beverage, that is, the one with chili chocolate, is not immediately to the left of the most expensive beverage.
7) The drink with Cointreau costs $7.
8) The cheapest drink is not Pine Nightcap.
9) The beverage with milk chocolate costs $1 more than the Comet Delight but $1 less than the drink with peppermint Schnapps.
10) The caramel chocolate is in a cup immediately to the left of a cup without candy cane and immediately to the right of the cup containing Kahlúa.
11) The beverage with Kahlúa does not cost $5.
12) The beverage containing rum is not immediately to the left of the cheapest drink.

	LIQUOR						CHOCOLATE TYPE						PRICE ($)					
	Amaretto	Baileys Irish Cream	Cointreau	Kahlúa	Peppermint Schnapps	Rum	Caramel	Chili	Dark	Milk	Orange	White	2	4	5	6	7	8
BEVERAGE																		
Comet Delight																		
Frosty Elixir																		
Mistletoe Mirage																		
Pine Nightcap																		
Snowfall Sip																		
Winter Dream																		
PRICE ($)																		
2																		
4																		
5																		
6																		
7																		
8																		
CHOCOLATE TYPE																		
Caramel																		
Chili																		
Dark																		
Milk																		
Orange																		
White																		

	Comet Delight	Frosty Elixir	Mistletoe Mirage	Pine Nightcap	Snowfall Sip	Winter Dream
LIQUOR						
HOCOLATE TYPE						
PRICE						

DECEMBER 6

SNOWY CREATIONS

Four children took part in a local competition to build the best snowman. Their creations, affectionately named Aspen, Claus, Ember, and Kelvin, stand proudly in the snow-covered square, showcasing the diverse talents of their creators.

A panel of judges meticulously examined the snowmen, assessing every detail of creativity and execution, and assigned points out of 100. However, no masterpiece is without its imperfections. Points were deducted for specific technical flaws, challenging the participants to refine their skills for future competitions.

Solve the puzzle to discover the names of the artists behind Aspen, Claus, Ember, and Kelvin, their scores, and the technical flaws pointed out by the judges. Left and right in the puzzle are from your point of view.

Clues

1) One snowman received a deduction because of the use of illegal snow (it was brought in from outside the allowed area). This snowman did not receive the highest score.
2) The snowman that received a score of 80 is somewhere to the right of the one built by Vera, but immediately to the left of the snowman that received a deduction for the use of illegal snow.
3) Nick's snowman is somewhere between the one rated 85/100 and the one penalized for the use of a fake carrot.
4) Sarah's snowman received a higher score than Claus but lower than the unstable snowman.
5) Kelvin, which was not rated 90/100, was not built by Gabby.

	Aspen	Claus	Ember	Kelvin
CREATOR	______	______	______	______
FLAW	______	______	______	______
SCORE	______	______	______	______

CREATORS: Gabby, Nick, Sarah, Vera.

FLAWS: Fake carrot, Illegal snow, Unfinished head, Unstable.

SCORES: 80, 85, 90, 95.

SLEIGH RIDE

After spending a pleasant afternoon adorning their sleighs with vibrant colors and funny names, five children are ready for a thrilling ride down the slope. As they make their way toward the hill, a strange coincidence catches their attention. Each child's jacket is one of the same five colors as the sleighs. However, no child is wearing a jacket that matches the color of his/her own sleigh. Solve the puzzle to determine which child owns each sleigh and match the correct colors with the corresponding jackets and sleighs.

Clues

1) Bridget, in a green jacket, is chatting with two friends: the child in the violet jacket and Caleb.
2) The jacket of the owner of the Arctic Express is the same color as the sleigh Snow Queen.
3) Daisy's sleigh is the same color as Caleb's jacket.
4) Felix's sleigh is not the same color as Daisy's jacket.
5) The Winter Wind, which is blue, is owned by the child in the white jacket.
6) Daisy owns the Polar Bullet.
7) George's sleigh is white.

		CHILD					SLEIGH COLOR					JACKET COLOR				
		Bridget	Caleb	Daisy	Felix	George	Blue	Green	Red	Violet	White	Blue	Green	Red	Violet	White
SLEIGH	Arctic Express															
	Glacial Glide															
	Polar Bullet															
	Snow Queen															
	Winter Wind															
JACKET COLOR	Blue															
	Green															
	Red															
	Violet															
	White															
SLEIGH COLOR	Blue															
	Green															
	Red															
	Violet															
	White															

SLEIGH	CHILD	SLEIGH COLOR	JACKET COLOR
Arctic Express			
Glacial Glide			
Polar Bullet			
Snow Queen			
Winter Wind			

SNOWY CREATIONS –
HINT
Start by figuring out who built Kelvin.

DECEMBER 7

CHRISTMAS IN VIENNA

In the enchanting city of Vienna, seven elderly friends on a Winter trip spent an entire day exploring the renowned Weinachtmarkt, immersing themselves in the festive atmosphere and savoring the season's flavors. After a day filled with laughter, warmth, and the aroma of freshly baked treats, they bought Christmas presents for their families back home and indulged in the city's culinary delights.

Solve the puzzle and uncover the present each person chose for their loved ones, the Viennese treats they tasted throughout the day, and the amounts they spent.

Clues

1) The person who ate Käsekrainer bought scented candles or handmade jewelry.
2) Of the person who ate the fluffy shredded pancakes called Kaiserschmarrn and the person who bought a woolen scarf, one spent €110.
3) The person who tasted Glühwein, a type of spice wine, spent 10 euros less than the person who bought handcrafted ornaments.
4) Henry spent 10 euros more than the person who ate Kaiserschmarrn.
5) The person who bought handcrafted ornaments did not eat Bratwurst.
6) The person who ate Lebkuchen spent 50 euros less than the person who bought a woolen scarf.
7) Beatrice spent less than the person who bought antique-style toys, who is not Stanley.
8) Of Eleanor and the person who ate Kaiserschmarrn, one spent €100 and the other bought antique-style toys.
9) The person who bought gingerbread cookies is either Henry or the person who ate Käsekrainer.
10) Of Walter and the person who spent €120, one bought the scented candle, and the other ate Sachertorte.
11) Eleanor, who did not eat Käsekrainer, spent 40 euros more than the person who bought a leather wallet.
12) Neither Francis, who ate Apfelstrudel, nor Henry bought the woolen scarf.

		PURCHASE							MONEY SPENT (€)							FOOD						
		Antique-style toys	Gingerbread cookies	Handcrafted ornaments	Handmade jewelry	Leather wallet	Scented candles	Woolen scarf	60	70	80	90	100	110	120	Apfelstrudel	Bratwurst	Glühwein	Kaiserschmarrn	Käsekrainer	Lebkuchen	Sachertorte
NAME	Beatrice																					
	Eleanor																					
	Francis																					
	Henry																					
	Margaret																					
	Stanley																					
	Walter																					
FOOD	Apfelstrudel																					
	Bratwurst																					
	Glühwein																					
	Kaiserschmarrn																					
	Käsekrainer																					
	Lebkuchen																					
	Sachertorte																					
MONEY SPENT (€)	60																					
	70																					
	80																					
	90																					
	100																					
	110																					
	120																					

NAME	PURCHASE	MONEY SPENT (€)	FOOD
Beatrice			
Eleanor			
Francis			
Henry			
Margaret			
Stanley			
Walter			

DECEMBER 8

ICE SKATING LESSONS

Last week, from Monday to Saturday, five sets of teenage siblings, each consisting of a brother and a sister, gathered at the local ice rink for a week-long skating experience. The girls, seasoned and graceful skaters, took it upon themselves to mentor their brothers, who were still beginners in the art of gliding on ice.

Each day, every girl provided one-on-one coaching to one of the boys. The arrangement was meticulous: no boy received lessons twice from the same girl throughout the week.

Use the clues to reconstruct the ice skating schedule. The second clue has already been entered into the table.

Clues

1) In every brother/sister pair, the two names start with the same letter.
2) Bianca's brother was taught by Fanny on Tuesday, by Elsa on Thursday, and by Diana on Friday.
3) Frederick was taught by Bianca on Tuesday, by Chloe on Thursday, and by Abigail on Friday.
4) On Monday, while Fanny was teaching her brother, Chloe was teaching the guy who skated on Tuesday with Diana and on Wednesday with Bianca.
5) Abigail skated with her brother on Wednesday, while Elsa skated with her brother on Friday.
6) On Wednesday, Elsa gave ice skating lessons to Calvin.
7) The guy who skated with Bianca on Friday also skated on Tuesday with Chloe and with his sister on Saturday.
8) On Thursday, Bianca gave ice skating lessons to the guy who then skated with his sister on Friday.

BROTHERS: Adam, Bernie, Calvin, Derek, Ethan, Frederick

TEACHER / DAY	Abigail	Bianca	Chloe	Diana	Elsa	Fanny
Monday						
Tuesday						Bernie
Wednesday						
Thursday					Bernie	
Friday				Bernie		
Saturday						

THE CHRISTMAS BOOK CLUB

At the last book club gathering, the members engaged in lively discussions about the perfect Christmas-themed book to delve into during the month of December. Use the clues to discover who championed each title, how many votes each book received, and their respective ratings on Goodreads.

Clues

1) The book with a rating of 4.07 received fewer votes than the one proposed by Taylor.
2) Jane suggested either "A Christmas Memory" or the book with a rating of 4.10.
3) The highest-rated book, proposed by Taylor but not "The Silver Skates," received fewer votes than the book proposed by Bi.
4) The book with the lowest rating, not "The Gift of the Magi," received two fewer votes than the book proposed by Jane but one fewer vote than "A Christmas Memory."
5) Of "A Christmas Memory" and the book that received 9 votes, one was suggested by Sasha, and the other has a rating of 4.24.
6) "The Gift of the Magi" received more votes than the book proposed by Taylor.
7) "A Christmas Carol" did not receive 8 or 9 votes.

		PROPOSED BY					RATING					VOTES				
		Bi	Jane	Lorentz	Sasha	Taylor	3.89	4.07	4.10	4.24	4.29	7	8	9	10	11
BOOK	A Christmas Carol															
	A Christmas Memory															
	Christmas Oranges															
	The Gift of the Magi															
	The Silver Skates															
VOTES	7															
	8															
	9															
	10															
	11															
RATING	3.89															
	4.07															
	4.10															
	4.24															
	4.29															

BOOK	PROPOSED BY	RATING	VOTES
A Christmas Carol			
A Christmas Memory			
Christmas Oranges			
The Gift of the Magi			
The Silver Skates			

DECEMBER 9

WINTER WARFARE

Every Christmas at the Johnson family estate, a peculiar tradition unfolds: the annual snowball battle, a grand event that unites all the Johnson children, cousins, and relatives. Laughter and excitement fill the air as the children split into two teams: The Arctic Avengers and the Snowflake Savages.

In the hours preceding the battle, the children designed their team emblems, forging a sense of camaraderie and unity. Now, as the sun sets, they prepare for the epic showdown. Each child has established their fortress in a different corner of the big garden, fortifying their positions with walls of snow and piling up snowballs for the initial barrage.

Each team member has an assigned role. A **scout** ventures into enemy territory, their eyes keen and watchful. A sharpshooting **sniper** stands guard, ensuring cover for the scout's risky escapades. An **engineer** is tasked with replenishing the team's snowball arsenal, providing a constant supply of ammunition. An **architect** stands ready to repair fort damages, maintaining the team's stronghold. Lastly, a **strategist** organizes the team's maneuvers. Last year, there was also a snowball medic providing first aid to the players knocked out during the battle. However, this year, the position has been abolished. Its very existence made it difficult to present the snowball battle as an innocuous pastime to the adults of the family.

Solve the puzzle to uncover the secrets of this winter battle: deduce each child's role, pinpoint their fort's location in the garden, and determine the precise number of snowballs they've amassed for the epic clash.

Clues on the Arctic Avengers

1) Of these three team members—Gabriel, the one who built a fort under the garden gazebo, and the one who built a fort next to the toolshed—the one who made the least amount of snowballs is the one who built the fort under the gazebo.
2) Logan, who made 54 snowballs, is either the sniper or the engineer.
3) The scout made fewer than 59 snowballs and did not build a fort among the fruit trees.
4) Owen is not the strategist of the team. Gabriel is neither the scout nor the sniper.
5) Neither the sniper nor the engineer built a fort on the vegetable patch.
6) One of the strategist and the engineer made 52 snowballs, and the other made 54.
7) The engineer built a fort next to the barbecue grill.
8) The sniper did not build a fort under the gazebo or among the fruit trees.
9) Maya built her fort either on the vegetable patch or beside the barbecue grill.

Clues on the Snowflake Savages

1) Of Lucas and Julian, one made 44 snowballs, and the other made 50.
2) Nathan did not build his fort next to the bird feeder.
3) The sniper is either Julian or Lucas.
4) The sniper made more snowballs than the team member who built a fort next to the apple tree but fewer snowballs than the team member who built a fort next to the garden shed.
5) The team member who built a fort under the swing set amassed more snowballs than the sniper.
6) The child who made 56 snowballs did not build a fort under the swing set or next to the bird feeder.
7) The strategist, whose name is not Lucas, did not build a fort behind the playhouse.
8) Of Samuel and Julian, one built a fort next to the garden shed.
9) Both the scout and the architect made more than 53 snowballs each.
10) The scout did not build a fort next to the bird feeder.
11) The engineer built a fort next to the garden shed.
12) The team member who made 54 snowballs is neither the sniper nor the scout.
13) The team member who made 54 snowballs did not build a fort next to the garden shed or the playhouse.

Linking Clue

The child who built a fort next to the playhouse made 6 fewer snowballs than the sniper of the Arctic Avengers.

		FORT LOCATION					ROLE					SNOWBALLS				
		Barbecue grill	Fruit trees	Garden gazebo	Toolshed	Vegetable patch	Architect	Engineer	Scout	Sniper	Strategist	52	54	56	58	60
CHILD	Claire															
	Gabriel															
	Logan															
	Maya															
	Owen															
SNOWBALLS	52															
	54															
	56															
	58															
	60															
ROLE	Architect															
	Engineer															
	Scout															
	Sniper															
	Strategist															

		FORT LOCATION					ROLE					SNOWBALLS				
		Apple tree	Bird feeder	Garden shed	Playhouse	Swing Set	Architect	Engineer	Scout	Sniper	Strategist	44	50	54	56	60
CHILD	Ella															
	Julian															
	Lucas															
	Nathan															
	Samuel															
SNOWBALLS	44															
	50															
	54															
	56															
	60															
ROLE	Architect															
	Engineer															
	Scout															
	Sniper															
	Strategist															

CHILD	FORT LOCATION	ROLE	SNOW-BALLS
Claire			
Gabriel			
Logan			
Maya			
Owen			

CHILD	FORT LOCATION	ROLE	SNOW-BALLS
Ella			
Julian			
Lucas			
Nathan			
Samuel			

DECEMBER 10

STOCKING STUFFER SURPRISE

It's Christmas morning, and excitement is palpable as four siblings bound out of bed to discover their stockings overflowing with surprises. Each child is overjoyed to find their favorite sweet treat tucked inside, along with a different accessory – gloves, socks, a hat, or a scarf. Using the clues below, can you determine what is inside each stocking and to which child it belongs? Left and right in the puzzle are from your point of view.

Clues

1) In the stocking with socks, there are nougat bars or chocolate coins.
2) Serge's stocking is next to the one containing socks.
3) Cam's stocking is not next to the one with the nougat bars.
4) The stocking with chocolate coins is somewhere to the left of the stocking containing a scarf but somewhere to the right of Martha's stocking.
5) The gloves and the marshmallows are in different stockings.
6) In stocking number 1 there are no chocolate coins or nougat bars.
7) The stocking with the hat is immediately left of Luisa's but immediately right of the stocking containing fruit gummies.
8) Martha's stocking does not contain fruit gummies.

CHILDREN: Cam, Luisa, Martha, Serge.
SWEETS: Chocolate coins, Fruit gummies, Marshmallows, Nougat bars.
ACCESSORIES: Gloves, Hat, Scarf, Socks.

TWINKLING TREES

The Todd children, excited about the upcoming Christmas festivities, have just finished decorating five magnificent trees in their garden. Each child decorated a different tree, choosing a unique light color and a distinct tree topper to add a personal touch. Using the following clues, can you determine the heights of the trees, identify the decorators, and match each tree with its specific light color and tree topper?

Clues

1) The lights on the tree topped with a star are neither red nor green.
2) The tree topped with a crown is taller than the one decorated with yellow lights.
3) The tree decorated with blue lights is 4 ft taller than the tree topped with a star.
4) Of the tree topped with an angel and the one decorated by Lauren, one is 24 ft tall, and the other one has yellow lights.
5) James decorated the tree 2 ft taller than the one with a star on top.
6) Amelia decorated the shortest tree.
7) The tree decorated by Lauren is taller than 23 ft.
8) Of Amelia and the child who topped the tree with a snowman, one used white lights. The other decorated the 26-foot-tall tree.
9) The five trees are the ones with white, red, and yellow lights, the tallest tree, and the one decorated by Gary.

		TREE TOPPER					LIGHT COLOR					HEIGHT (FT)				
		Angel	Bow	Crown	Snowman	Star	Blue	Green	Red	White	Yellow	20	22	24	26	28
SIBLING	Amelia															
	Gary															
	James															
	Lauren															
	Megan															
HEIGHT (FT)	20															
	22															
	24															
	26															
	28															
SLEIGH COLOR	Blue															
	Green															
	Red															
	White															
	Yellow															

STOCKING STUFFER SURPRISE - HINT

Start by figuring out which accessory is contained in stocking number 1.

SIBLING	TREE TOPPER	LIGHT COLOR	HEIGHT (FT)
Amelia			
Gary			
James			
Lauren			
Megan			

DECEMBER 11

PRESENTS BIG AND SMALL

Santa Claus meticulously prepared gifts for six families, ensuring a large package and a small one for each household. The beautifully wrapped six pairs of packages are neatly arranged on two shelves, waiting to bring joy to the families. Using the clues provided, can you match each family with their specific pair of packages and discover the surprises hidden within?

Clues

1) The delivery for the Hudson family, which does not include the painting kit, is numbered one higher than the delivery that includes a whiteboard (which is in a box without a bow).
2) The packages for the Carter family are either immediately above or immediately below those for the Hudsons.
3) The gifts for the Millers include a portable speaker in a box without a bow.
4) The delivery for the Reeds is numbered twice that for the Lees.
5) The delivery with a smartwatch is immediately above or immediately below the one for the Lees.
6) The two boxes containing a science book and an activity dome are either both with a bow or both without.
7) Delivery number 5 does not include a science book.
8) The interactive globe and the painting kit will go to different families.
9) The space projector is not for the Lees or the Reeds.
10) The pair of packages with the space projector is neither immediately above nor immediately below the pair with the painting kit.
11) The delivery for the Fosters is numbered one lower than the delivery containing the gaming desk, which is in a box without a bow.
12) The gaming desk is in the delivery immediately above or immediately below the one with the LEGO set.

		BIG BOX						SMALL BOX						FAMILY					
		Activity dome	Gaming desk	Interactive globe	Pop up tent	Space projector	Whiteboard	LEGO set	Musical toy	Painting kit	Portable speaker	Science book	Smartwatch	Carter	Foster	Hudson	Lee	Miller	Reed
NUMBER	1																		
	2																		
	3																		
	4																		
	5																		
	6																		
FAMILY	Carter																		
	Foster																		
	Hudson																		
	Lee																		
	Miller																		
	Reed																		
SMALL BOX	LEGO set																		
	Musical toy																		
	Painting kit																		
	Portable speaker																		
	Science book																		
	Smartwatch																		

BIG BOX ______ ______ ______

SMALL BOX ______ ______ ______

FAMILY ______ ______ ______

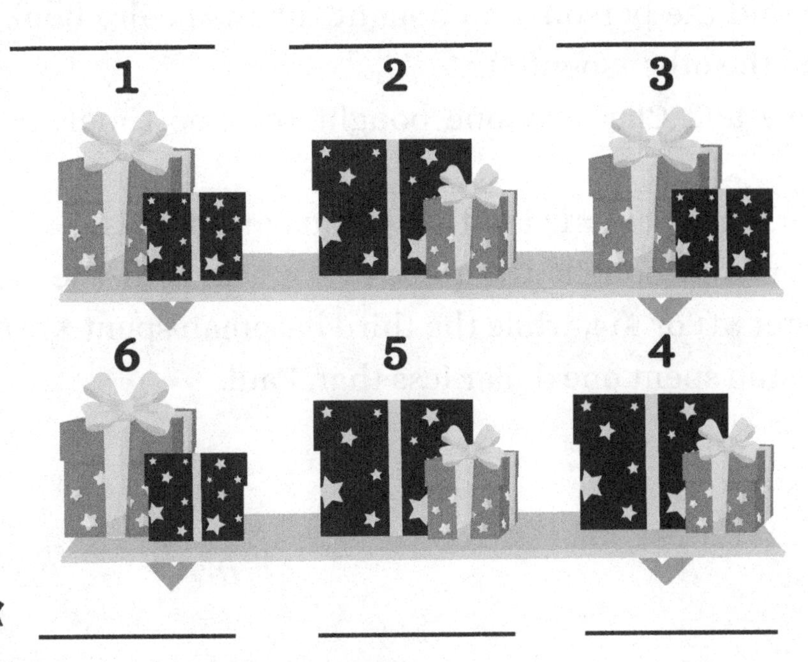

BIG BOX ______ ______ ______

SMALL BOX ______ ______ ______

FAMILY ______ ______ ______

DECEMBER 12

THE GRAND SLAM GIFT EXCHANGE

In the spirit of camaraderie, seven members of a baseball team organized a secret Santa gift exchange. Each player bought a present within the budget of 10 to 20 dollars. Use the clues to discover the chosen presents, their prices, the players who bought them, and their respective positions on the team.

Clues

1) The first baseman did not buy the holy water hip flask or the socks with the writing "Bring Me Wine."

2) The first baseman spent $2 more than the person who bought the nose hair trimmer.

3) The holy water hip flask, not bought by Joshua, cost one dollar more than the poo timer.

4) Clarence spent one dollar less than the player who bought the nose hair trimmer.

5) The socks with the "Bring Me Wine" writing were bought by Matthew or Paul.

6) Steven plays as pitcher or third baseman.

7) The outfielder, not Robert, spent one dollar more than the person who bought the poo timer.

8) Of Steven and the person who bought the insult flip book, one plays as second baseman, and the other spent $12.

9) Of Matthew and Clarence, one bought the poo timer, and the other plays as catcher.

10) The player who spent $13 is either Joshua or Thomas.

11) The player who bought the pink eye mask with the writing "The Queen is Sleeping" spent $11 or $14, while the third baseman spent $12 or $15.

12) The shortstop spent one dollar less than Paul.

		GIFT BOUGHT							PRICE ($)							POSITION						
		Beard lights	Bring Me Wine socks	Holy water hip flask	Insult flip book	Nosehair trimmer	Poo timer	The Queen is Sleeping eye mask	11	12	13	14	15	16	17	First baseman	Second baseman	Third baseman	Catcher	Outfielder	Pitcher	Shortstop
NAME	Clarence																					
	Joshua																					
	Matthew																					
	Paul																					
	Robert																					
	Steven																					
	Thomas																					
POSITION	First baseman																					
	Second baseman																					
	Third baseman																					
	Catcher																					
	Outfielder																					
	Pitcher																					
	Shortstop																					
PRICE ($)	11																					
	12																					
	13																					
	14																					
	15																					
	16																					
	17																					

NAME	GIFT BOUGHT	PRICE ($)	POSITION
Clarence			
Joshua			
Matthew			
Paul			
Robert			
Steven			
Thomas			

DECEMBER 13

CHRISTMAS CLAUS-TROPHOBIA

This Christmas, Little Tommy is in for a magical surprise. Several of his relatives had the brilliant idea of dressing up as Santa Claus to shower him with presents. As he eagerly awaits the gift-giving, he finds himself surrounded by four Santas, each bearing a different number of gifts. Use the clues to identify who is hiding behind each costume and how many gifts they brought for Tommy.

Clues

1) Santa number 2 carries more gifts than Eugene, but fewer than Tommy's grandfather.
2) Tommy's father has fewer presents in his sack than Dennis but more than Santa number 4
3) Santa number 1 is Joseph.
4) Tommy's uncle is immediately between Eugene and the Santa with 11 gifts in his sack.
5) Bradley is Tommy's cousin.

NAMES: Bradley, Dennis, Eugene, Joseph.

RELATIONS: Cousin, Father, Grandfather, Uncle.

PRESENTS: 5, 7, 10, 11.

SECRET SANTA RELOADED

Five colleagues from the office—three men named Andrew, Carl, and Erik, and two women named Betty and Daniela—decided to organize a Secret Santa exchange. Each person contributed a small gift, forming a delightful assortment of presents. The colleagues gathered to exchange gifts on the last day before the Christmas holiday. Each person randomly picked one item from the contributions made by the others. Use the clues to match each colleague with the gift they contributed, the person who picked their gift, and the pattern of wrapping paper used for each present.

Clues

1) Daniela contributed a mug to the Secret Santa. She did not wrap it in damask paper.

2) No two gifters and receivers form a pair of swappers. In other words, if A picked the gift contributed by B, then B did not pick the item contributed by A.

3) Betty was delighted to receive candles. Her contributed gift was not a Christmas towel.

4) The notebook is wrapped in retro paper.

5) Carl wrapped his gift, not the fancy chocolate bar, in striped paper.

6) Erik picked the gift contributed by Andrew. This was the only instance of the giver and receiver being of the same sex.

7) Andrew picked the gift in the floral package.

8) A man picked the towel.

		RECEIVER					GIFT					PAPER PATTERN				
		Andrew	Betty	Carl	Daniela	Erik	Candles	Chocolate	Mug	Notebook	Towel	Damask	Floral	Polka dots	Retro	Stripes
GIVER	Andrew															
	Betty															
	Carl															
	Daniela															
	Erik															
PAPER PATTERN	Damask															
	Floral															
	Polka dots															
	Retro															
	Stripes															
GIFT	Candles															
	Chocolate															
	Mug															
	Notebook															
	Towel															

GIVER	RECEIVER	GIFT	PAPER PATTERN
Andrew			
Betty			
Carl			
Daniela			
Erik			

DECEMBER 14

THE BAKERSHOP QUARTET

by Bethany Nuckolls

The bake case at Fuba's coffee house always features delicious treats, including an assortment of gingerbread men, made in house, and each decorated uniquely, with a different neck accessory, three or four buttons, and either a smile or a frown, so that no two are exactly the same. Using the clues below, can you recreate the gingerbread men in the bake case? Left and right are from your point of view.

Clues

1) Two of the gingerbread men are decorated with scarves, but they are not next to each other, nor do they have the same number of buttons.
2) No gingerbread man with three buttons is frowning.
3) The gingerbread man in position C is wearing a necktie and four buttons.
4) The only frowning gingerbread man is to the left of the one decorated with a bowtie, who has three buttons.

	A	B	C	D
ACCESSORY				
BUTTONS				
EXPRESSION				

THE ENCHANTED EVERGREENS

Six creative individuals participated in a Christmas tree decorating competition. Each adorned tree was displayed throughout the day, inviting people to admire the intricate designs and cast their votes. Can you determine the name of the decorator behind each tree, the peculiarity of their decoration, and the number of votes received by each tree?

Clues

1) Michelle decorated the tree immediately between the one with sustainable decor and the one that received 16 votes, which has a star on top.
2) Kenneth's tree received 11 votes, while tree number 2 received the most votes.
3) Neither Kenneth nor Susan used origami ornaments.
4) Jason, who used artificial snow, decorated the tree numbered half of the one that received 30 votes. Cheryl did not put a star on the top of her tree.
5) The decorations on tree number 5 involve some tech elements.
6) The tree with edible ornaments, which was not decorated by Susan, received 48 votes.
7) Neither tree 1 nor tree 6 has edible ornaments.
8) Tree number 1 is not the one that received 24 votes.

		PECULIARITY						DECORATOR						VOTES					
		Artificial snow	Edible ornaments	Glowing globes	Origami ornaments	Sustainable decor	Tech elements	Cheryl	Frank	Jason	Kenneth	Michelle	Susan	11	16	24	30	48	51
NUMBER	1																		
	2																		
	3																		
	4																		
	5																		
	6																		
VOTES	11																		
	16																		
	24																		
	30																		
	48																		
	51																		
DECORATOR	Cheryl																		
	Frank																		
	Jason																		
	Kenneth																		
	Michelle																		
	Susan																		

ECULIARITY ______ ______ ______ ______ ______ ______

ECORATOR ______ ______ ______ ______ ______ ______

VOTES ______ ______ ______ ______ ______ ______

DECEMBER 15

GRANDMA'S MERRY MYSTERY

Grandma Elsa eagerly welcomed her ten grandchildren, forming five pairs of brothers and sisters, to celebrate the start of the Christmas holiday. However, before they could indulge in Grandma Elsa's delightful cuisine, the children faced a strenuous logical challenge. Grandma's living room had a big decagonal table with ten seating spots numbered 1 to 10. Grandma Elsa presented the children with a set of clues, challenging them to figure out the unique seating arrangement that would fulfill all her requirements. Can you help them crack the puzzle and find their designated spots at the table? You should also be informed that no brother has a name starting with the same letter as his sister's. Moreover, remember that Denise's last name is Baker and that Teresa is not Albert's sister.

Clues

1) The five sisters will occupy the even-numbered seats, while the brothers will sit in the odd-numbered seats.
2) Scott will be seated at number 1. His sister will not occupy seat number 6.
3) Seat 2 will be taken by a girl whose last name is Vega.
4) The boy with the last name Nash will not be seated at spot number 3.
5) Teresa's brother will be seated at number 9.
6) No boy will sit next to his sister.
7) Immediately to Shirley's right, there will be Albert.
8) Albert's sister will not sit at spot number 10.
9) The girl with last name Austin will occupy a seat with a number higher than Shirley's but lower than Harold's spot number.
10) Albert's seat number is higher than the Baker boy's.
11) Dylan will have Denise seated immediately to his left and Harold's sister seated immediately to his right.
12) Dylan will be seated at a spot whose number is half that of his sister's.

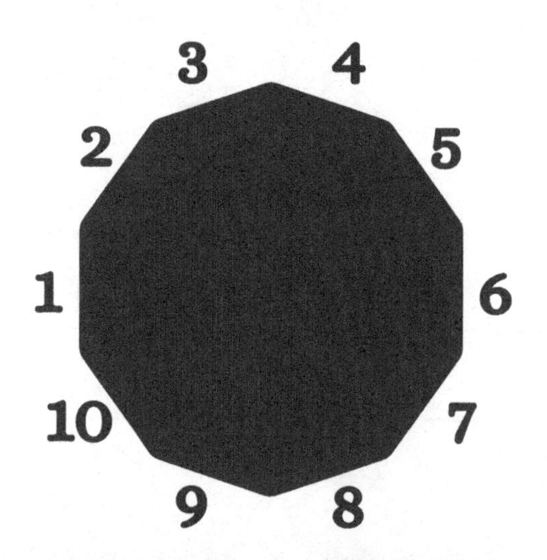

		BROTHER'S NAME					BROTHER'S SEAT					SISTER'S NAME					SISTER'S SEAT				
		Albert	Dylan	Harold	Scott	Tyler	1	3	5	7	9	Alexis	Denise	Helen	Shirley	Teresa	2	4	6	8	10
LAST NAME	Austin																				
	Baker																				
	Nash																				
	Robinson																				
	Vega																				
SISTER'S SEAT	2																				
	4																				
	6																				
	8																				
	10																				
SISTER'S NAME	Alexis																				
	Denise																				
	Helen																				
	Shirley																				
	Teresa																				
BROTHER'S SEAT	1																				
	3																				
	5																				
	7																				
	9																				

LAST NAME	BROTHER'S NAME	BROTHER'S SEAT	SISTER'S NAME	SISTER'S SEAT
Austin				
Baker				
Nash				
Robinson				
Vega				

DECEMBER 16

SANTA'S TOYBOX PUZZLE

Santa finds himself in a whirlwind of preparations. He urgently needs six old-school toys and six cuddly stuffed animals to complete his special gifts. The toys and stuffed animals are stored in the six boxes in the picture. It's up to the diligent elves to decipher the clues and determine the precise position of each toy and each stuffed animal. Can you help them?

Clues

1) The stuffed penguin is in the box directly above the box containing the kaleidoscope.
2) The stuffed penguin is in the box immediately to the left and on the same shelf as the box containing the jump rope.
3) The box containing the stuffed giraffe has a number twice that of the box containing the Yo-Yo, which is directly above the box with the stuffed monkey.
4) The box containing the rubber ball is numbered two lower than the box with the stuffed dolphin, which is not on the same row as the box containing a set of marbles.
5) The box containing the wooden whistle is on the same shelf and somewhere to the left of the box with the stuffed koala.

OLD-FASHIONED TOYS: Jump rope, Kaleidoscope, Marbles, Rubber ball, Whistle, Yo-Yo.

STUFFED ANIMALS: Dolphin, Giraffe, Koala, Monkey, Penguin, Sloth.

OLD-FASHIONED TOY ________ ________ ________

STUFFED ANIMAL ________ ________ ________

1 2 3

4 5 6

OLD-FASHIONED TOY ________ ________ ________

STUFFED ANIMAL ________ ________ ________

HOLIDAYS AT GRANDMA'S

by Bethany Nuckolls

Ethel's seven grandchildren are over for the holidays, and the busy grandmother is finding it rather trying to keep such young, energetic kids occupied. Fortunately, she knows the secret to any grandchild's heart—home-baked cookies. Armed with cookie cutters, three different colors of icing, and a variety of baking decorations, Ethel is planning on having a splendid winter's afternoon with her decidedly creative progeny. As each child created a unique cookie (no two made the exact same cookie when considering the combination of shape, color, and decoration), can you determine the specifications of each holiday cookie?

Clues

1) Neither Janice nor Sephora decorated a tree cookie.
2) No two stars were frosted the same color.
3) Exactly two bell cookies were decorated, but not by Curtis.
4) May and Timothy both used sprinkles and decorated a star and a bell in some order.
5) No one who used gold dragées decorated a star.
6) The three grandchildren who used colored sugar each used a different color of frosting.
7) The only child who frosted a green tree used sprinkles.
8) If Curtis used red frosting, then Vera used white frosting. Otherwise, Curtis used white frosting and so did Timothy.
9) If Janice used red frosting, then she also used sprinkles. Otherwise, she used green frosting and so did Rory.
10) If Rory decorated a star, then Timothy did not. Otherwise, Rory decorated a tree with colored sugar.

COOKIE SHAPES: Bell, Star, Tree.

FROSTING COLORS: Green, Red, White.

DECORATIONS: Colored sugar, Gold dragées, Rainbow sprinkles.

CHILD	COOKIE SHAPE	FROSTING COLOR	DECORATION
Curtis			
Janice			
May			
Rory			
Sephora			
Timothy			
Vera			

DECEMBER 17

MERRY CARD-MAS

In the festive season, Wendy finds herself overseeing her five children as they craft personalized Christmas cards for various family members. Each child has chosen a distinct style for their card, ranging from elegant embossed designs to playful pop-up creations and heartfelt photo cards. Adding another layer of uniqueness, each card showcases a different image, capturing the essence of the holiday spirit.

Your task is to solve the puzzle and determine the recipient of each child's card, the specific type of card they chose, the featured image, and the age of each young artist.

Clues

1) The embossed Christmas card does not feature candles and was not created by the 13-year-old sibling.
2) The card with candles is not intended for Grandpa Jim.
3) Carter is the author of the 3D card.
4) The glittery card was crafted by the 9-year-old sibling.
5) Julian made the card for Aunt Sharon.
6) The photo card unmistakably displays a family picture.
7) Sadie created the card depicting a snowman, while Avery made the one showing a snowy landscape.
8) Julian is two years younger than the sibling who made the photo card.
9) The sibling who made the card for Grandma Izzy is either two years older or two years younger than the one who created the embossed card.
10) Sadie is two years older than the sibling who made the card for Uncle Rick but two years younger than the one who created the card featuring Santa Claus.
11) Avery is two years older than the sibling who made the card for Aunt Amanda.
12) The child who crafted the pop-up card is two years older than Lyla.

		RECIPIENT					SIBLING'S AGE					IMAGE					CARD TYPE				
		Aunt Amanda	Aunt Sharon	Grandma Izzy	Grandpa Jim	Uncle Rick	7	9	11	13	15	Candles	Family photo	Landscape	Santa Claus	Snowman	3D Card	Embossed	Glittery	Photo	Pop-up
SIBLING	Avery																				
	Carter																				
	Lyla																				
	Julian																				
	Sadie																				
CARD TYPE	3D Card																				
	Embossed																				
	Glittery																				
	Photo																				
	Pop-up																				
IMAGE	Candles																				
	Family photo																				
	Landscape																				
	Santa Claus																				
	Snowman																				
SIBLING'S AGE	7																				
	9																				
	11																				
	13																				
	15																				

SIBLING	RECIPIENT	SIBLING'S AGE	IMAGE	CARD TYPE
Avery				
Carter				
Lyla				
Julian				
Sadie				

DECEMBER 18

THE UGLY SWEATER SOIREE

Rita, a Christmas enthusiast, adores the holiday season, especially for her annual tradition of hosting an Ugly Sweater Party. This year, the theme is tacky Christmas sweaters. Each guest must don the most outrageously designed sweater they can find. To add to the merriment, Rita decided to spice things up by having each guest organize a unique Christmas-themed game for the party. In the spirit of giving, she also encouraged her friends to bring presents to share with others.

Rita's guest list comprised seven of her closest friends. Your challenge is to use the provided clues to match each friend with their specific Christmas sweater, the game they organized, and the number of presents they brought to the party.

Clues

1) The guest wearing a sweater representing a flashy Christmas tree brought two more presents than the guest who organized the game of Pass the Ornament.
2) Kingsley brought four more presents than the guest who organized the game of Pass the Ornament.
3) Viviana brought five more presents than the guest who organized the game of Christmas Jeopardy.
4) The guest wearing a sweater representing a drunk elf brought fewer gifts than the person whose sweater depicts a melted snowman.
5) The person who brought ten gifts does not wear a sweater depicting a melted snowman.
6) Jayden brought three fewer presents than the person who organized the game of Christmas Pictionary.
7) Either Nicole or the guest who brought 9 gifts organized the game of Christmas Scattergories. The other wore the truck driver Santa sweater.
8) Corey, the guest who organized the game of Tinsel Limbo, and the one wearing a toothless reindeer sweater are three different guests.
9) The guest with the pooping reindeer sweater is either the one who brought 11 presents or the one who organized the Christmas movie trivia game.
10) Sheldon brought two more gifts than the guest in the truck driver Santa sweater.
11) Corey brought one fewer gift than the guest in the drunk elf sweater.
12) The guest who organized the candy cane hunt brought three more gifts than the guest in the naked Santa sweater.
13) The seven guests are Corey, Jayden, the two who organized the games of Tinsel Limbo and Christmas Pictionary, the one in the truck driver Santa sweater, and the two who brought 6 and 9 gifts.

		GAME							SWEATER							GIFTS						
		Candy cane hunt	Christmas Jeopardy	Christmas Pictionary	Christmas movie trivia	Christmas Scattergories	Pass the ornament	Tinsel limbo	Drunk elf	Flashy Christmas tree	Melted snowman	Naked Santa	Pooping reindeer	Toothless reindeer	Truck driver Santa	5	6	7	8	9	10	11
GUEST	Corey																					
	Deegan																					
	Jayden																					
	Kinglesy																					
	Nicole																					
	Sheldon																					
	Viviana																					
GIFTS	5																					
	6																					
	7																					
	8																					
	9																					
	10																					
	11																					
SWEATER	Drunk elf																					
	Flashy Christmas tree																					
	Melted snowman																					
	Naked Santa																					
	Pooping reindeer																					
	Toothless reindeer																					
	Truck driver Santa																					

GUEST	GAME	SWEATER	GIFTS
Corey			
Deegan			
Jayden			
Kingelsy			
Nicole			
Sheldon			
Viviana			

DECEMBER 19

SPICE GIRLS

by Bethany Nuckolls

It's another day at Fuba's coffee shop, and five new gingerbread men are displayed in the bake case... or rather, gingerbread women! Each has been decorated uniquely so that no two are exactly identical. Can you recreate these holiday treats using the clues below?

Clues

1) Each gingerbread woman has either curly or straight hair and there is at least one instance of each. Each gingerbread woman has either a striped, polka-dotted, or checked scarf, and there is at least one instance of each. And finally, each gingerbread woman is wearing either mittens or boots, and there is at least one instance of each.

2) No two gingerbread women with boots are immediately next to each other. At least two gingerbread women wear boots.

3) At least one gingerbread woman has curly hair and a checked scarf, but not the one in position D.

4) Neither the gingerbread woman in position A, nor the gingerbread woman in position E, sports polka-dotted scarves.

5) Three gingerbread women have straight hair, but not the one in position C.

6) If the gingerbread woman in position D has a checked scarf, then so does the gingerbread woman in position B, and vice versa. Otherwise, they are both wearing striped scarves.

7) No gingerbread woman is wearing both a striped scarf and boots.

8) If the gingerbread woman in position C has nothing in common with the one in position D, then she also has nothing in common with the one in position B. Otherwise, she has at least one thing in common with the one in position D and at least one thing in common with the one in position B.

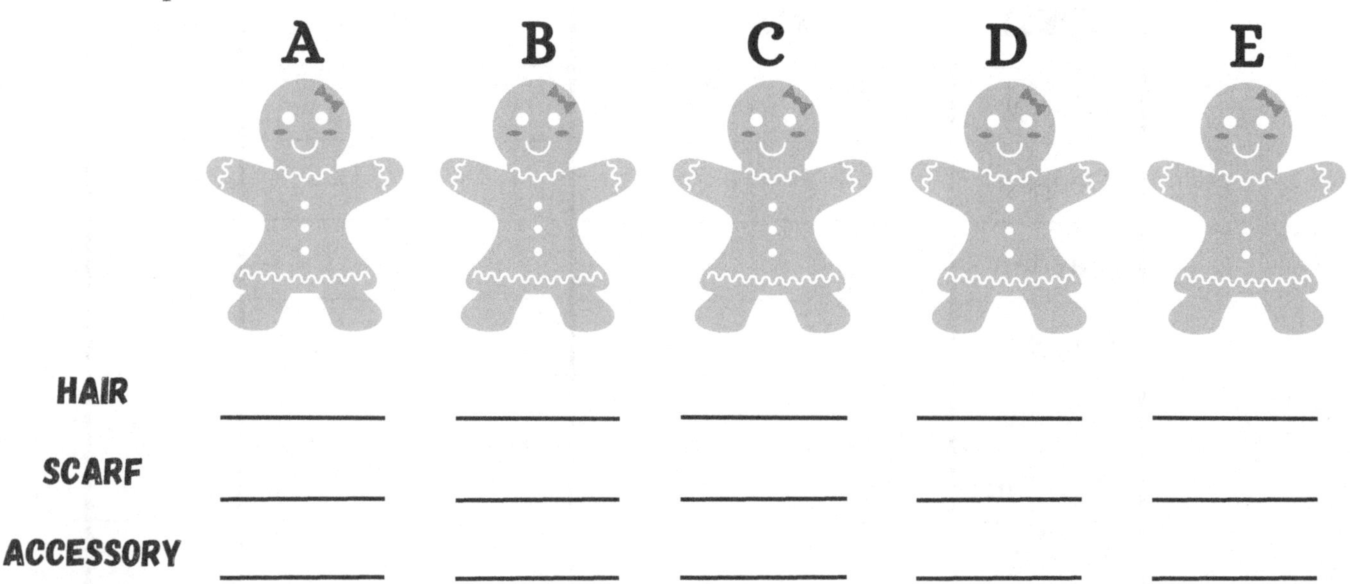

BITE-SIZED WONDERS

In the Gingerbread House Creation Challenge, five teams showcased their creativity. Each team, with a different number of members, chose a unique theme for their gingerbread house. Moreover, they incorporated special edible elements into their creations. Solve the puzzle to uncover the details of each team's masterpiece.

Clues

1) Of The Ginger Group and Sugar Squad, one team comprises 14 members, and the other is the team that constructed a house featuring licorice ladders.
2) Frosty Fanatics is either the team consisting of 14 members or the team that created a traditional gingerbread house.
3) Sweet Artisans has one fewer member than the team that used marshmallow snow.
4) The team that crafted a fairy tale-inspired gingerbread house has three fewer members than Frosty Fanatics.
5) The team Icing Innovators has one more member compared to the team that incorporated chocolate roof tiles into their design.
6) One more person built the gingerbread castle than the fairy tale-inspired gingerbread house.
7) The Sweet Artisans adorned their creation with candy cane columns.
8) The team that incorporated chocolate roof tiles into their design consists of one more member than the team that built the space-inspired gingerbread house.
9) The Sugar Squad did not use marshmallow snow in their gingerbread house design.

		ELEMENT					THEME					TEAM SIZE				
		Candy cane columns	Chocolate roof tiles	Licorice ladders	Marshmallow snow	Pretzel fencing	Castle	Fairy tale	Space	Traditional	Under the sea	12	13	14	15	16
TEAM NAME	Frosty Fanatics															
	Icing Innovators															
	Sugar Squad															
	Sweet Artisans															
	The Ginger Group															
TEAM SIZE	12															
	13															
	14															
	15															
	16															
THEME	Castle															
	Fairy tale															
	Space															
	Traditional															
	Under the sea															

TEAM NAME	ELEMENT	THEME	TEAM SIZE
Frosty Fanatics			
Icing Innovators			
Sugar Squad			
Sweet Artisans			
The Ginger Group			

DECEMBER 20

THE ORNAMENT CHALLENGE

At the local school, two boys, Cyrus and Stanley, along with three girls, Alicia, Eva, and Mollie, have just completed their Christmas project. Each child painted a Christmas ornament and is now eagerly awaiting the chance to showcase their masterpiece on the family tree.

The ornaments are made of different materials, adding a delightful variety to the holiday decorations. Use the provided clues to determine the material of each child's ornament and the color they chose. Left and right are from your point of view.

Clues

1) Both ornament A and ornament D were painted by girls.
2) The blue ornament, not E, is made neither of leather nor ceramic.
3) The ornament painted by Cyrus is not made of plastic.
4) Mollie's ornament is somewhere to the right of the leather ornament but somewhere to the left of the red ornament, which is made of glass.
5) Stanley did not paint his ceramic ornament white.
6) The white ornament is not made of wood.
7) Alicia's yellow ornament is not made of plastic or wood.
8) Eva's ornament is more than one place to the right of the green ornament.
9) Ornament B is not made of ceramic or leather.

CHILDREN: Alicia, Cyrus, Eva, Mollie, Stanley.

COLORS: Blue, Green, Red, White, Yellow.

MATERIALS: Ceramic, Glass, Leather, Plastic, Wood.

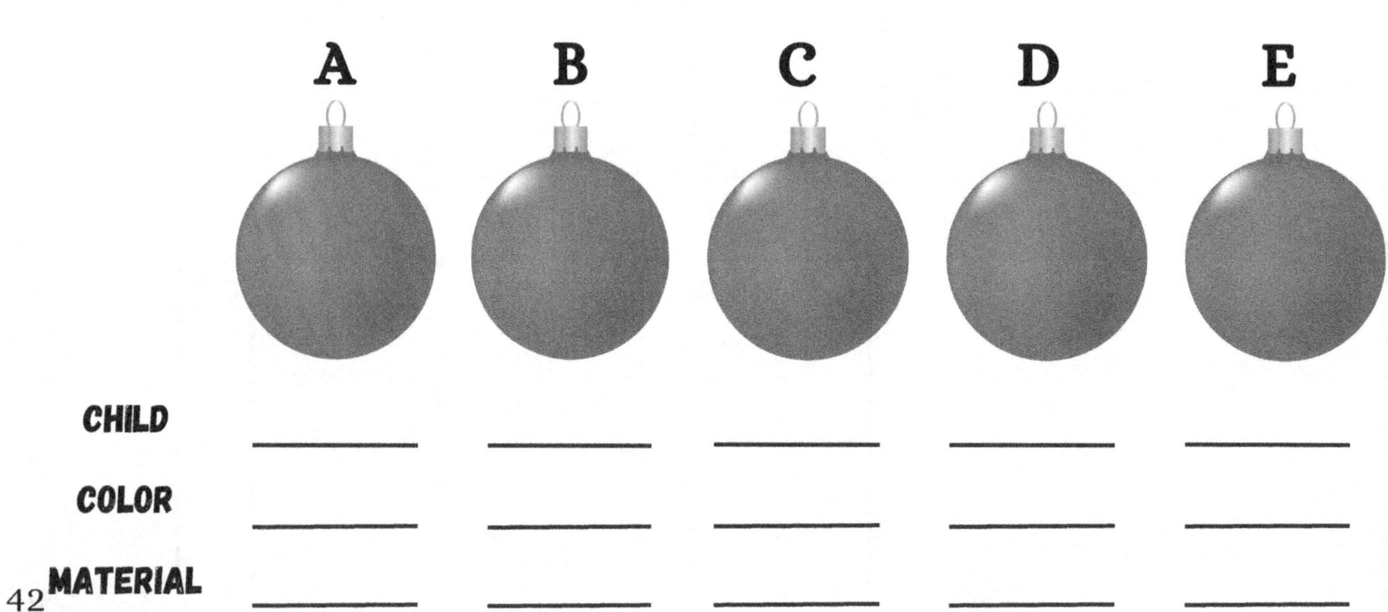

FESTIVE FACADES

In Butterfly Road, five families in houses 1 to 5 transformed their front yards uniquely for Christmas. Solve the puzzle to discover their distinctive decorations and the day each family chose to adorn their homes.

Clues

1) Of the Dixon family and the family who put up their decorations on December 17, one lives at number 2 and the other used ornaments.

2) The family that adorned their front yard with lanterns was not the first to set up decorations. Additionally, they do not live at number 2.

3) Of the Young family and the family at number 2, one used pinecones and the other set up decorations on December 7.

4) Neither the Alvarez family (who does not live at number 5) nor the family that used a wreath to decorate their house resides at number 1.

5) Either the Alvarez family or the family that decorated last used lanterns, and the other used a wreath.

6) The family at number 5 set up decorations sometime after the family at number 3 but sometime before the Halls.

7) The family residing at number 1 did not decorate on December 7.

		DECORATIONS					HOUSE NUMBER					DAY				
		Garlands	Lanterns	Ornaments	Pinecones	Wreath	1	2	3	4	5	December 2	December 7	December 12	December 17	December 22
FAMILY	Alvarez															
	Dixon															
	Hall															
	Lynch															
	Young															
DAY	December 2															
	December 7															
	December 12															
	December 17															
	December 22															
HOUSE NUMBER	1															
	2															
	3															
	4															
	5															

FAMILY	DECORATIONS	HOUSE NUMBER	DAY
Alvarez			
Dixon			
Hall			
Lynch			
Young			

DECEMBER 21

A HOLIDAY HOMECOMING

The Samuelson family's six lively youngsters are at home for the holiday season. With the upcoming week in mind, they've come up with a special plan. Each day, starting from Monday and ending on Saturday, they've scheduled a different activity to enjoy together. Each day's activity will be supervised by a different family member.

Solve the puzzle to determine which child suggested each activity, which day it's planned for, and which adult family member will be supervising the fun-filled day.

Clues

1) The day the children will spend with their father will come sometime after ice skating but sometimes before the Christmas lights scavenger hunt.
2) Ice skating was not the activity proposed by Albert.
3) The children will not go ice skating with their grandfather.
4) The day in which the children will have a Christmas movie marathon is not Wednesday.
5) Grandpa will not be involved in Monday's activity.
6) Tess proposed either ice skating or the activity the children will do with their uncle.
7) On Wednesday, the children will not spend the day with their uncle.
8) One of Jean and Gary proposed the Friday activity, and the other suggested the activity with their aunt.
9) The Christmas lights scavenger hunt will not occur the day before the board games.
10) The children will not play board games on Tuesday.
11) The children will not make Christmas crafts on Wednesday.
12) One of Ella and Albert proposed the Monday activity, and the other suggested making Christmas crafts.
13) The six activities are going Christmas shopping, playing board games, the activity supervised by grandma, the activities planned for Monday and Wednesday, and the one proposed by Gary.
14) Seth proposed watching Christmas movies or the Tuesday activity, or perhaps both.
15) The children will not watch Christmas movies with their uncle.

		ACTIVITY						PROPOSED BY						SUPERVISED BY					
		Board games	Christmas crafts	Christmas lights hunt	Christmas movies	Christmas shopping	Ice skating	Albert	Ella	Gary	Jean	Seth	Tess	Aunt	Father	Grandma	Grandpa	Mother	Uncle
DAY	Monday																		
	Tuesday																		
	Wednesday																		
	Thursday																		
	Friday																		
	Saturday																		
SUPERVISED BY	Aunt																		
	Father																		
	Grandma																		
	Grandpa																		
	Mother																		
	Uncle																		
PROPOSED BY	Albert																		
	Ella																		
	Gary																		
	Jean																		
	Seth																		
	Tess																		

DAY	ACTIVITY	PROPOSED BY	SUPERVISED BY
Monday			
Tuesday			
Wednesday			
Thursday			
Friday			
Saturday			

DECEMBER 22

SCENTED SECRETS

Grandma Caterina has set up a festive display of Christmas candles, casting a warm glow throughout the room. Each candle is numbered from 1 to 9. Within the arrangement, some candles are tall, while others are shorter. Your task is to use the provided clues to deduce the scent emanating from each candle. Please keep in mind that left and right are referenced from your perspective.

Clues

1) The lilac and eucalyptus candles are of the same height.
2) The lilac-scented candle is positioned somewhere to the right of both the rosemary candle and the sage-scented candle.
3) The lime-scented candle, a tall one, is positioned somewhere to the left of the vanilla-scented candle, which is immediately to the right of a short candle.
4) The lilac-scented candle is numbered twice as the bergamot-scented candle, which is somewhere to the left of the short tangerine-scented candle.
5) The tangerine-scented candle is not adjacent to the rosemary-scented candle, which has an odd number and is positioned somewhere to the right of the cinnamon-scented candle.
6) The cinnamon-scented candle is somewhere to the left of the sage-scented candle, which is immediately next to the bergamot candle.
7) The bergamot-scented candle is somewhere to the left of the vanilla-scented candle.

SCENTS: Bergamot, Cinnamon, Eucalyptus, Lilac, Lime Rosemary, Sage, Tangerine, Vanill

SCENT _____ _____ _____ _____ _____ _____ _____ _____ _____

HIKE & SEEK

During their mountain vacation, James and his parents explored different trails each day. James, with his sharp eyes, spotted various animals on their hikes. Solve the puzzle to uncover which trail they took each day, its length, and the animal James spotted on each path.

Clues

1) James spotted the red fox the day after he hiked the Sapphire Summit but the day before he tackled the 4-mile hike.

2) Neither the 3-mile hike nor the one where James spotted an elk took place on Thursday. Moreover, neither occurred on Frostbite Ridge.

3) The Ember Valley hike was completed the day before the hike in which James spotted the owl. Neither of these hikes happened on Wednesday.

4) The shortest hike occurred the day after James spotted an elk.

5) On Avalanche Pass, James spotted either an elk or an owl.

6) Both the trail where James saw an elk and the 3-mile one were completed before Frostbite Ridge.

7) Thunderbolt Walk was completed the day before James saw a mountain lion. Both events occurred on days later than the 7-mile hike.

8) James saw a marmot the day before he walked the 5-mile trail.

9) The hike taken on Wednesday was 6 miles long.

		TRAIL					MILES					ANIMAL				
		Avalanche Pass	Ember Valley	Frostbite Ridge	Sapphire Summit	Thunderbolt Walk	3	4	5	6	7	Elk	Marmot	Mountain lion	Owl	Red fox
DAY	Monday															
	Tuesday															
	Wednesday															
	Thursday															
	Friday															
ANIMAL	Elk															
	Marmot															
	Mountain lion															
	Owl															
	Red fox															
MILES	3															
	4															
	5															
	6															
	7															

DAY	TRAIL	MILES	ANIMAL
Monday			
Tuesday			
Wednesday			
Thursday			
Friday			

DECEMBER 23

WE THE PEOPLE

by Bethany Nuckolls

Fuba's coffee house is at it again. Since their gingerbread men and women have become so popular, this morning they prepared 7 new creations in the bake case—the Gingerbread People! Avoiding gendered clothing this time, the bakers frosted each gingerbread with a single scarf patterned in a different combination of three colors. Using the clues below, can you determine the colors on each gingerperson's scarf? (Note: Cookies are labeled A through G, from left to right. Left and right always refer to your perspective.)

Clues

1) The gingerbread person in position D is wearing a scarf with the color blue.
2) The gingerbread people in positions E and G share two of the same colors on their scarves.
3) The gingerbread person is position A isn't wearing any white.
4) The pink-purple-blue scarf is worn by a gingerbread person which is displayed somewhere to the right of the one decorated with the gray-white-purple scarf.
5) The gingerbread person sporting the orange-white-pink scarf is not immediately beside the gingerbread person wearing the green-white-blue scarf.
6) No two gingerbread people decorated with pink icing are immediately next to each other.

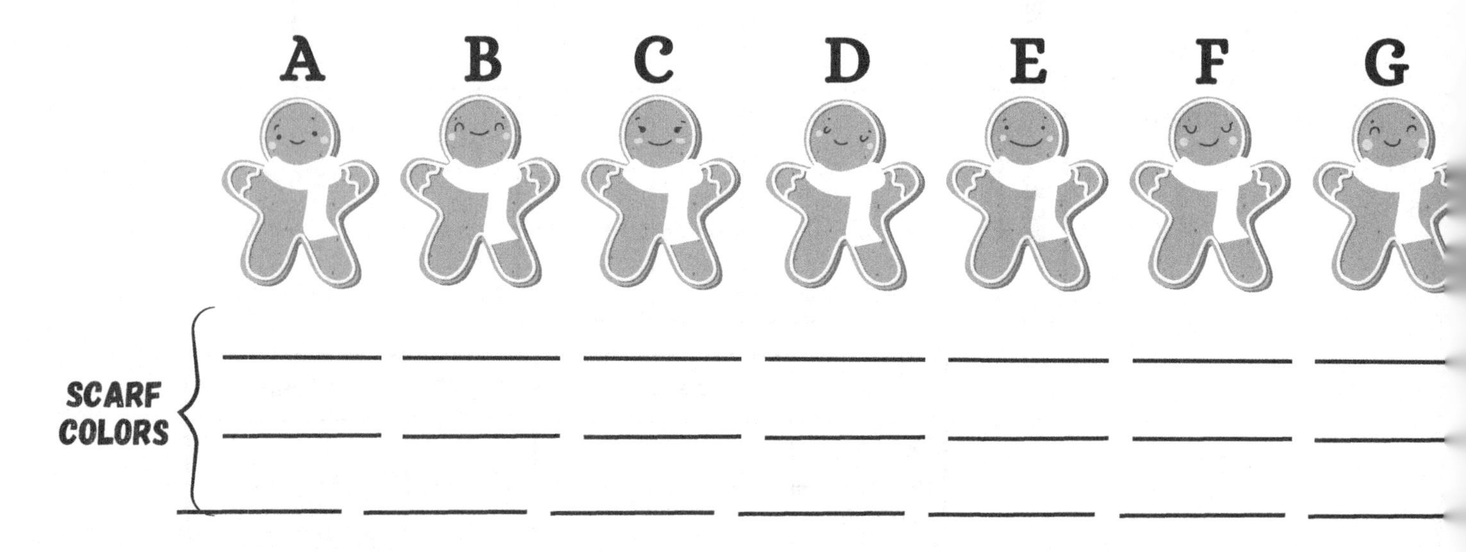

SCARF COLOR COMBINATIONS: Blue-pink-white, Gray-white-purple, Green-white-blue, Orange-white-pink, Pink-purple-blue, Pink-yellow-blue, Yellow-white-purple.

SEASONAL STALLS

At a bustling Christmas market, five sellers occupy neighboring stands, showcasing their unique homemade products. Solve the puzzle to dentify the specific products each person sells, their stand numbers, and the ways in which they adorned their stands.

Clues

1) Neither the soap vendor nor Caleb decorated their stand with seasonal foliage.
2) The pottery vendor, who adorned their stand with a Christmas tablecloth, is in a higher-numbered stand than the person who used paper snowflakes as decoration
3) Of Elena and the person who used fairy lights as decoration, one sells handcrafted ornaments, and the other is in stand number 4.
4) Of the person selling hand-poured candles and the one who used fairy lights as decoration, one is Nathan and the other is in stand 2.
5) The stand with paper snowflakes is a higher-numbered stand than the one decorated with garlands.
6) Of Olga and Caleb, one sells soap, and the other is in stand 5.
7) Olga is not in stand 1. Additionally, she did not decorate her stand with fairy lights.
8) The seller in stand 2 did not use seasonal foliage to decorate.
9) The soap vendor did not adorn their stand with paper snowflakes.

		SELLER					STAND					DECORATIONS				
		Caleb	Elena	Nathan	Olga	Steph	1	2	3	4	5	Garlands	Fairy lights	Seasonal foliage	Snowflakes	Tablecloths
PRODUCT	Artisanal soap															
	Handcrafted ornaments															
	Hand-poured candles															
	Hand-knit items															
	Handmade pottery															
DECORATIONS	Garlands															
	Fairy lights															
	Seasonal foliage															
	Snowflakes															
	Tablecloth															
STAND	1															
	2															
	3															
	4															
	5															

PRODUCT	SELLER	STAND	DECORATIONS
Artisanal soap			
Handcrafted ornaments			
Hand-poured candles			
Hand-knit items			
Handmade pottery			

DECEMBER 24

ASPEN'S LIFT LINE ENIGMA

Six buddies are enjoying a ski trip in Aspen and they are curently waiting in line for the chair lift to take them to the top of the slope. Each friend has purchased a ski pass loaded with 30 rides. Use the provided clues to figure out the name of each friend in the line and how many rides remain on their ski pass.

Clues

1) Elliot is immediately behind Morgan but immediately ahead of the skier with 13 rides on their ski pass.
2) Reuben has three fewer rides left on his ski pass than the person immediately behind him.
3) Arthur has one fewer ride left than Elliot but two more rides than Isaac.
4) Isaac is immediately behind Lacey but immediately ahead of the skier with 12 rides left on the ski pass.
5) The skier with 15 rides left is somewhere ahead of the skier with 14 rides left on the ski pass.

NAMES: Arthur, Elliot, Isaac, Lacey, Morgan, Reuben.

RIDES: 12, 13, 14, 15, 16, 17.

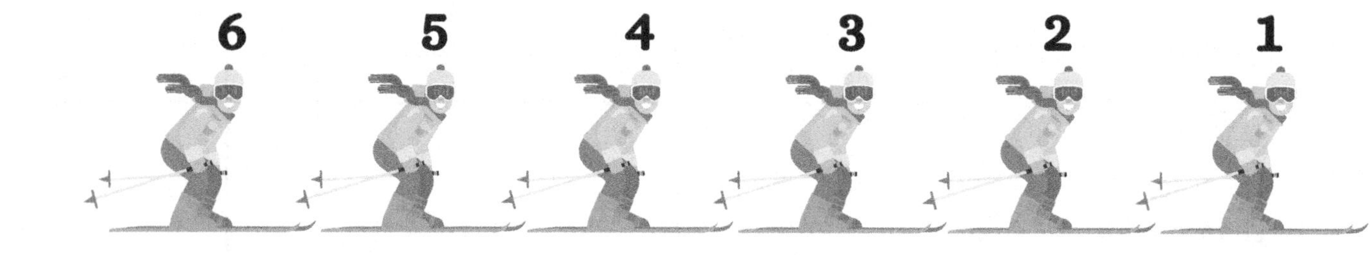

NAME ________ ________ ________ ________ ________ ________

RIDES ________ ________ ________ ________ ________ ________

BAKING UP HOLIDAY FUN

Five children gathered for a delightful Christmas activity - baking sugar cookies together. Each child chose a unique cookie shape and used a distinct color of icing sugar to decorate their creations. Use the clues to figure out how many cookies each child baked and the age of each young baker.

Clues

1) Isabel baked one fewer cookie than the child who baked candy cane-shaped cookies.
2) Luke is 3 years younger and baked 2 fewer cookies than the child who used blue icing sugar.
3) The white icing sugar was not used on the bell-shaped cookies.

4) The cookies with the white icing sugar are not 18 in number.

5) One of Yasmin and Devon made the star-shaped cookies, and the other used blue icing sugar.

6) The child who made the bell-shaped cookies is three years older than the child who used red icing sugar, either Isabel or Luke.

7) The child who made the bell-shaped cookies is younger than the child who baked the most cookies.

8) Devon's cookies are neither in the shape of bells nor snowflakes.

9) The child who made the snowman cookies is two years younger than the child who used yellow icing sugar.

10) The youngest child is either Devon or Yasmin.

11) Mike used either white or green icing sugar.

		SHAPE					COLOR					AGE					COOKIES				
		Bell	Candy cane	Snowflake	Snowman	Star	Blue	Green	Red	White	Yellow	6	7	9	10	12	17	18	20	21	23
CHILD	Devon																				
	Isabel																				
	Luke																				
	Mike																				
	Yasmin																				
COOKIES	17																				
	18																				
	20																				
	21																				
	23																				
AGE	6																				
	7																				
	9																				
	10																				
	12																				
COLOR	Blue																				
	Green																				
	Red																				
	White																				
	Yellow																				

CHILD	SHAPE	COLOR	AGE	COOKIES
Devon				
Isabel				
Luke				
Mike				
Yasmin				

DECEMBER 25

THE GREAT CHRISTMAS CHALLENGE

It is a crisp and snowy Christmas day, and in the cozy home of the Greens, the spirit of the festive season fills the air. This year, the family is about to journey down memory lane. Seven family members have brought collections of photo albums, each capturing a different Christmas from the 1980s. Gathered around the fireplace, the family members reminisce about the good old days, sifting through the photographs depicting their past Christmas celebrations. Each person brought a different number of photo albums, ranging from 2 to 9. The snapshots were taken in seven consecutive years from the '80s. Use the clues to determine the year in which each set of pictures was taken, the activities the family engaged in, the dishes they savored for lunch, and the board games that brought joy to their Christmases.

Clues

1) The pictures showing the family baking Christmas cookies are not the oldest ones.
2) There is an odd number of photo albums from 1984.
3) More years are separating the pictures of the family baking cookies and the pictures of the lunch with the stuffed pork loin than separating the pictures of the seafood paella and the pictures of the stuffed pork loin.
4) There are more than two photo albums from the year the family played Risk. It was not the same year in which all the family wrote letters to Santa.
5) The most recent set of pictures and the one showing the family playing Ticket to Ride both consist of an odd number of albums.
6) Aunt Patricia brought fewer photo albums than the person whose pictures show the family playing in the snow.
7) The year the family played in the snow, the main course was not lobster thermidor.
8) Cousin Linda's pictures do not show the family playing Scrabble.
9) The year the family ate seafood paella, they did not play Scrabble.
10) The pictures showing the lunch with the stuffed pork loin are the fourth oldest.
11) There are at least two more photo albums from the Christmas lunch with the seafood paella than that with the stuffed pork loin.
12) There are exactly three photo albums from the year in which the entire family read Christmas stories together in front of the chimney. These are not the pictures brought by Aunt Patricia.
13) Of the year immediately before the year the family played in the snow and the year immediately after it, one is documented in an even number of albums and the other in an odd number of albums.
14) The pictures brought by Grandma Evelyn are from the year before the epic game of Ticket to Ride, but they are more recent than those showing a game of Risk.
15) Adding together the number of albums from the two years the family played Risk and Ticket to Ride, one obtains the number of albums brought by Grandma Evelyn.
16) The person who brought the most photo albums, not Grandpa Victor, has at least two more albums than those from the year the family ate baked ham.
17) In the best documented Christmas (the one with the largest number of photo albums) the family played Pictionary.
18) The year in which a magnificent scavenger hunt was organized, the Christmas lunch had no baked ham.
19) Grandpa Victor did not bring the pictures showing the lunch with the baked ham.
20) The pictures in Aunt Patricia's albums are at least three years older than those in the collection of six albums.
21) The collection of six albums is not the most recent.
22) Both the pictures brought by Aunt Mary and those showing the family enjoying a roast lamb were taken later than 1985.
23) In 1985, the family created Christmas crafts together and enjoyed prime rib roast for lunch. Of this year, there are three fewer photo albums than of the year in which the family played Monopoly.
24) The seven picture collections are the one with seven albums, those brought by Uncle Max (which does not consist of exactly 4 albums) and Uncle Richard, the pictures from 1983, the pictures showing the family playing Dixit (not from 1988), the pictures of the tree decoration process, and those of the lunch with the Beef Wellington (not taken in 1984).
25) The collection of six albums shows a lunch with lobster termidor but does not show the family playing Uno.
26) The photos in the six-album collection are not those taken the year before the pictures in Cousin Linda's albums.

	ACTIVITY	YEAR	MAIN COURSE	PHOTO ALBUMS	BOARD GAME
	Bake Christmas cookies Create Christmas crafts Decorate the tree Play in the snow Read Christmas stories Scavenger hunt Write letters to Santa		Baked ham Beef Wellington Lobster thermidor Prime rib roast Roast lamb Seafood paella Stuffed pork loin		Dixit Monopoly Pictionary Risk Scrabble Ticket to Ride Uno
RELATIVE Aunt Mary Aunt Patricia Cousin Linda Grandma Evelyn Grandpa Victor Uncle Max Uncle Richard					
BOARD GAME Dixit Monopoly Pictionary Risk Scrabble Ticket to Ride Uno					
PHOTO ALBUMS					
MAIN COURSE Baked ham Beef Wellington Lobster thermidor Prime rib roast Roast lamb Seafood paella Stuffed pork loin					
YEAR					

RELATIVE	ACTIVITY	YEAR	MAIN COURSE	PHOTO ALBUMS	BOARD GAME
Aunt Mary					
Aunt Patricia					
Cousin Linda					
Grandma Evelyn					
Grandpa Victor					
Uncle Max					
Uncle Richard					

YOUR FREE GIFT

Ignite your puzzle-solving superpowers with this complete guide

SOLUTIONS

DECEMBER 1

THE CHRISTMAS PRANK

1	2	3	4	5	6	7	8	9
Zoe	Ginny	Chris	John	Lia	Mary	Bob	Ruth	Kyle

THE NORTH POLE SLEIGH RACE

REINDEER	ELF	SLEIGH COLOR	PLACEMENT	DECORATION
Blitzen	Jingles	Red	4	Ornaments
Comet	Twinkle	White	2	Holly
Dancer	Pudding	Gold	1	Lights
Rudolph	Sparkle	Blue	5	Bells
Vixen	Merry	Silver	3	Ribbons

DECEMBER 2

HO-HO-HO AND WHO LIVES BELOW

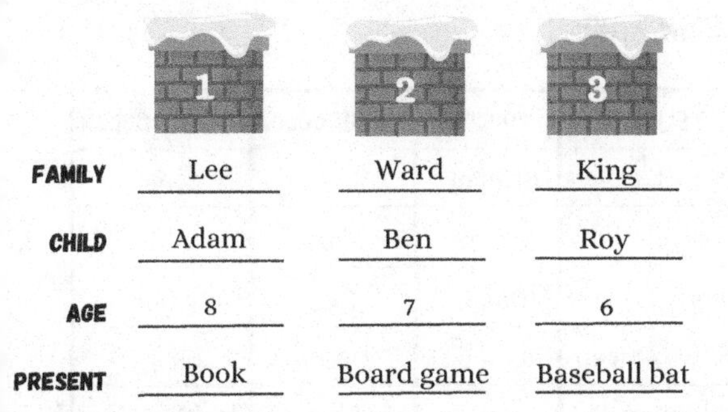

	1	2	3
FAMILY	Lee	Ward	King
CHILD	Adam	Ben	Roy
AGE	8	7	6
PRESENT	Book	Board game	Baseball bat

CAROLS AND CHARITY

FAMILY	SONG	SWEETS	DONATION ($)
Brown	Feliz Navidad	Toffees	50
Garcia	Joy to the World	Pecan pie bars	40
Miller	Silent Night	Eggnog fudge	30
Smith	O Holy Night	Rum balls	60
Wilson	Jingle Bells	Cookies	70

DECEMBER 3

THE MERRY HOSTS

HOST	CHRISTMAS MOVIE	DAY	SNACK
Alice	The Polar Express	Tuesday	Stuffed mushrooms
Jack	Arthur Christmas	Monday	Peppermint popcorn
Katherine	The Grinch	Wednesday	Cinnamon pretzels
Nora	Elf	Thursday	Gingerbread cookies
Rachel	Home Alone	Sunday	Cranberry brie bites
Sam	Jingle All the Way	Saturday	Mini meatballs
Trevor	Rise of the Guardians	Friday	Fruit skewers

DECEMBER 4

ELVES UNSCRAMBLED

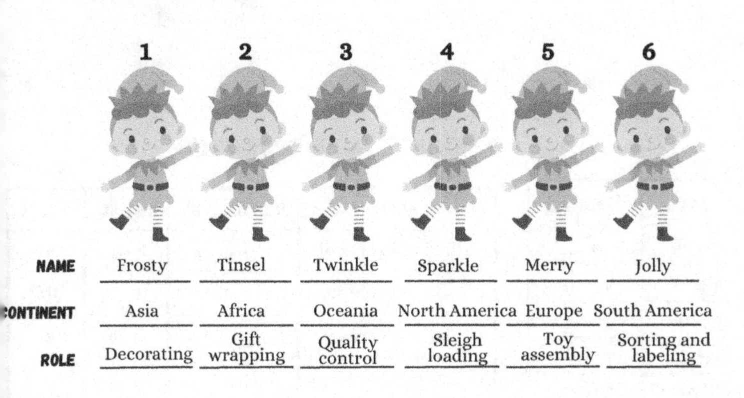

	1	2	3	4	5	6
NAME	Frosty	Tinsel	Twinkle	Sparkle	Merry	Jolly
CONTINENT	Asia	Africa	Oceania	North America	Europe	South America
ROLE	Decorating	Gift wrapping	Quality control	Sleigh loading	Toy assembly	Sorting and labeling

DECORATIVE DILEMMAS

SIBLING	ORNAMENTS	COLOR	NUMBER
Alex	Globes	Blue	24
Bella	Angels	Red	34
Chad	Figurines	Green	29
Luis	Eggs	Yellow	44
Naomi	Icicles	Silver	39

DECEMBER 5

HOLIDAY HOTCHOCO HEIST

	Comet Delight	Frosty Elixir	Mistletoe Mirage	Pine Nightcap	Snowfall Sip	Winter Dream
LIQUOR	Rum	Peppermint Schnapps	Kahlúa	Baileys Irish Cream	Amaretto	Cointreau
CHOCOLATE TYPE	Orange	Dark	Chili	Caramel	White	Milk
PRICE	$6	$8	$2	$5	$4	$7

DECEMBER 6

SNOWY CREATIONS

	Aspen	Claus	Ember	Kelvin
CREATOR	Vera	Gabby	Nick	Sarah
FLAW	Unstable	Fake carrot	Illegal snow	Unfinished head
SCORE	95	80	90	85

SLEIGH RIDE

SLEIGH	CHILD	SLEIGH COLOR	JACKET COLOR
Arctic Express	Bridget	Violet	Green
Glacial Glide	George	White	Blue
Polar Bullet	Daisy	Red	Violet
Snow Queen	Caleb	Green	Red
Winter Wind	Felix	Blue	White

DECEMBER 7

CHRISTMAS IN VIENNA

NAME	PURCHASE	MONEY SPENT (€)	FOOD
Beatrice	Leather wallet	60	Lebkuchen
Eleanor	Handmade jewelry	100	Bratwurst
Francis	Handcrafted ornaments	90	Apfelstrudel
Henry	Gingerbread cookies	80	Glühwein
Margaret	Antique-style toys	70	Kaiserschmarrn
Stanley	Scented candles	120	Käsekrainer
Walter	Woolen scarf	110	Sachertorte

DECEMBER 8

ICE SKATING LESSONS

DAY \ TEACHER	Abigail	Bianca	Chloe	Diana	Elsa	Fanny
Monday	Bernie	Adam	Ethan	Calvin	Derek	Frederick
Tuesday	Calvin	Frederick	Derek	Ethan	Adam	Bernie
Wednesday	Adam	Ethan	Bernie	Frederick	Calvin	Derek
Thursday	Derek	Calvin	Frederick	Adam	Bernie	Ethan
Friday	Frederick	Derek	Calvin	Bernie	Ethan	Adam
Saturday	Ethan	Bernie	Adam	Derek	Frederick	Calvin

THE CHRISTMAS BOOK CLUB

BOOK	PROPOSED BY	RATING	VOTES
A Christmas Carol	Lorentz	4.07	7
A Christmas Memory	Bi	4.24	10
Christmas Oranges	Taylor	4.29	8
The Gift of the Magi	Jane	4.10	11
The Silver Skates	Sasha	3.89	9

DECEMBER 9

WINTER WARFARE

CHILD	FORT LOCATION	ROLE	SNOW-BALLS
Claire	Garden gazebo	Strategist	52
Gabriel	Fruit trees	Architect	60
Logan	Barbecue grill	Engineer	54
Maya	Vegetable patch	Scout	58
Owen	Toolshed	Sniper	56

CHILD	FORT LOCATION	ROLE	SNOW-BALLS
Ella	Bird feeder	Architect	54
Julian	Apple tree	Strategist	44
Lucas	Playhouse	Sniper	50
Nathan	Swing set	Scout	60
Samuel	Garden shed	Engineer	56

DECEMBER 10

STOCKING STUFFER SURPRISE

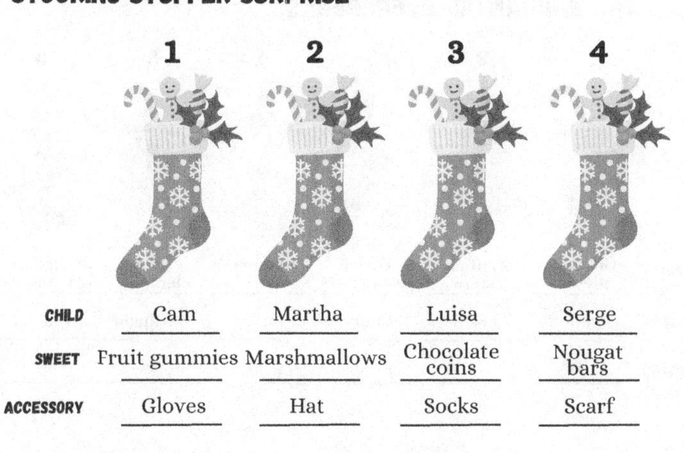

	1	2	3	4
CHILD	Cam	Martha	Luisa	Serge
SWEET	Fruit gummies	Marshmallows	Chocolate coins	Nougat bars
ACCESSORY	Gloves	Hat	Socks	Scarf

TWINKLING TREES

SIBLING	TREE TOPPER	LIGHT COLOR	HEIGHT (FT)
Amelia	Star	White	20
Gary	Angel	Blue	24
James	Bow	Red	22
Lauren	Snowman	Yellow	26
Megan	Crown	Green	28

DECEMBER 11

PRESENTS BIG AND SMALL

	1	2	3
BIG BOX	Space projector	Activity dome	Pop up tent
SMALL BOX	Portable speaker	LEGO set	Painting kit
FAMILY	Miller	Carter	Lee

	6	5	4
BIG BOX	Interactive globe	Gaming desk	Whiteboard
SMALL BOX	Science book	Musical toy	Smartwatch
FAMILY	Reed	Hudson	Foster

DECEMBER 12

THE GRAND SLAM GIFT EXCHANGE

NAME	GIFT BOUGHT	PRICE ($)	POSITION
Clarence	The Queen is Sleeping eye mask	11	Catcher
Joshua	Insult flip book	13	Second baseman
Matthew	Poo timer	14	First baseman
Paul	Bring Me Wine socks	17	Pitcher
Robert	Beard lights	16	Shortstop
Steven	Nose hair trimmer	12	Third baseman
Thomas	Holy water hip flask	15	Outfielder

DECEMBER 13

CHRISTMAS CLAUS-TROPHOBIA

	1	2	3	4
NAME	Joseph	Dennis	Eugene	Bradley
RELATION	Grandfather	Uncle	Father	Cousin
PRESENTS	11	10	7	5

SECRET SANTA RELOADED

GIVER	RECEIVER	GIFT	PAPER PATTERN
Andrew	Erik	Towel	Damask
Betty	Andrew	Chocolate	Floral
Carl	Betty	Candles	Stripes
Daniela	Carl	Mug	Polka dots
Erik	Daniela	Notebook	Retro

DECEMBER 14

THE BAKERSHOP QUARTET

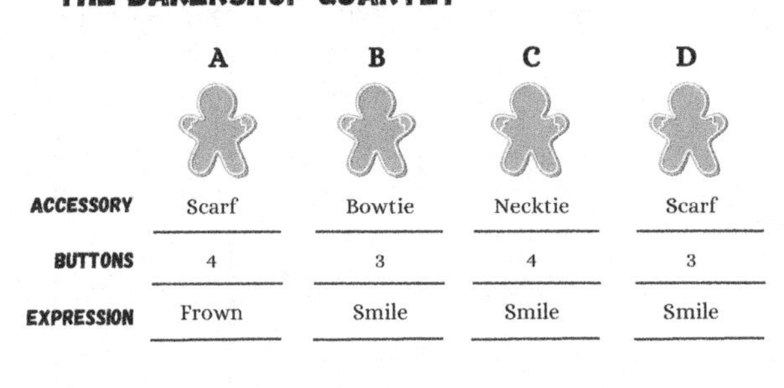

	A	B	C	D
ACCESSORY	Scarf	Bowtie	Necktie	Scarf
BUTTONS	4	3	4	3
EXPRESSION	Frown	Smile	Smile	Smile

THE ENCHANTED EVERGREENS

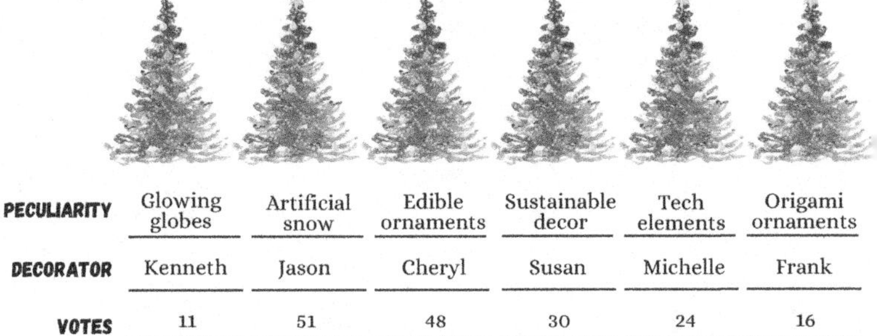

	1	2	3	4	5	6
PECULIARITY	Glowing globes	Artificial snow	Edible ornaments	Sustainable decor	Tech elements	Origami ornaments
DECORATOR	Kenneth	Jason	Cheryl	Susan	Michelle	Frank
VOTES	11	51	48	30	24	16

DECEMBER 15

GRANDMA'S MERRY MYSTERY

LAST NAME	BROTHER'S NAME	BROTHER'S SEAT	SISTER'S NAME	SISTER'S SEAT
Austin	Albert	5	Helen	8
Baker	Scott	1	Denise	4
Nash	Tyler	7	Alexis	10
Robinson	Dylan	3	Shirley	6
Vega	Harold	9	Teresa	2

DECEMBER 16

SANTA'S TOYBOX PUZZLE

	1	2	3
OLD-FASHIONED TOY	Rubber ball	Yo-Yo	Jump rope
STUFFED ANIMAL	Sloth	Penguin	Dolphin

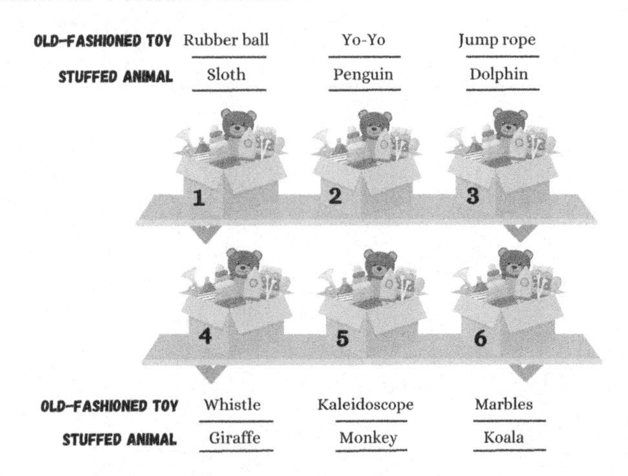

	4	5	6
OLD-FASHIONED TOY	Whistle	Kaleidoscope	Marbles
STUFFED ANIMAL	Giraffe	Monkey	Koala

HOLIDAYS AT GRANDMA'S

CHILD	COOKIE SHAPE	FROSTING COLOR	DECORATION
Curtis	Tree	White	Colored sugar
Janice	Bell	Green	Gold dragées
May	Star	White	Rainbow sprinkles
Rory	Star	Green	Colored sugar
Sephora	Star	Red	Colored sugar
Timothy	Bell	White	Rainbow sprinkles
Vera	Tree	Green	Rainbow sprinkles

DECEMBER 17

MERRY CARD-MAS

SIBLING	RECIPIENT	SIBLING'S AGE	IMAGE	CARD TYPE
Avery	Grandpa Jim	15	Landscape	Pop-up
Carter	Uncle Rick	7	Candles	3D Card
Lyla	Aunt Amanda	13	Family photo	Photo
Julian	Aunt Sharon	11	Santa Claus	Embossed
Sadie	Grandma Izzy	9	Snowman	Glittery

DECEMBER 18

THE UGLY SWEATER SOIREE

GUEST	GAME	SWEATER	GIFTS
Corey	Pass the ornament	Naked Santa	5
Deegan	Christmas Jeopardy	Drunk elf	6
Jayden	Christmas movie trivia	Flashy Christmas tree	7
Kingelsy	Christmas Scattergories	Melted snowman	9
Nicole	Candy cane hunt	Truck driver Santa	8
Sheldon	Christmas Pictionary	Toothless reindeer	10
Viviana	Tinsel limbo	Pooping reindeer	11

DECEMBER 19

SPICE GIRLS

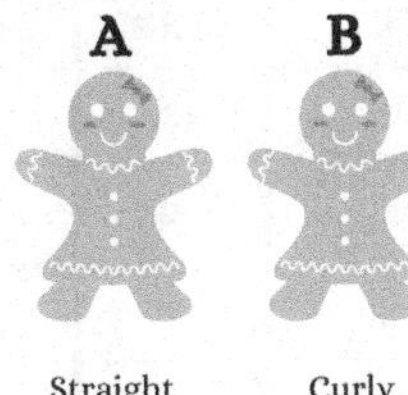

	A	B	C	D	E
HAIR	Straight	Curly	Curly	Straight	Straight
SCARF	Striped	Checked	Polka-dotted	Checked	Checked
ACCESSORY	Mittens	Boots	Mittens	Mittens	Boots

BITE-SIZED WONDERS

TEAM NAME	ELEMENT	THEME	TEAM SIZE
Frosty Fanatics	Chocolate roof tiles	Traditional	15
Icing Innovators	Pretzel fencing	Under the sea	16
Sugar Squad	Licorice ladders	Fairy tale	12
Sweet Artisans	Candy cane columns	Castle	13
The Ginger Group	Marshmallow snow	Space	14

DECEMBER 20

THE ORNAMENT CHALLENGE

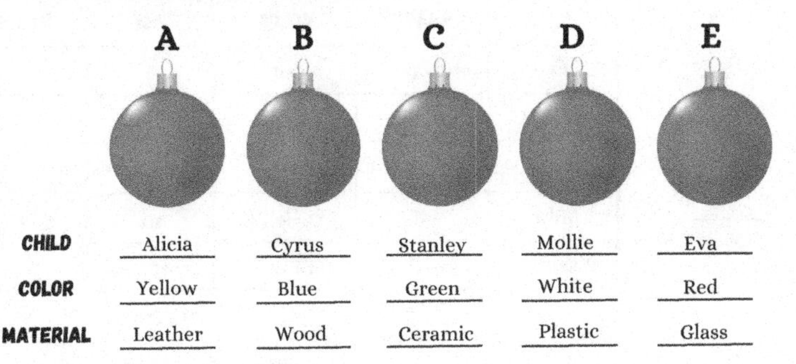

	A	B	C	D	E
CHILD	Alicia	Cyrus	Stanley	Mollie	Eva
COLOR	Yellow	Blue	Green	White	Red
MATERIAL	Leather	Wood	Ceramic	Plastic	Glass

FESTIVE FACADES

FAMILY	DECORATION	HOUSE NUMBER	DAY
Alvarez	Lanterns	3	December 12
Dixon	Garlands	2	December 7
Hall	Wreath	4	December 22
Lynch	Ornaments	5	December 17
Young	Pinecones	1	December 2

DECEMBER 21

A HOLIDAY HOMECOMING

DAY	ACTIVITY	PROPOSED BY	SUPERVISED BY
Monday	Ice skating	Ella	Mother
Tuesday	Christmas shopping	Seth	Father
Wednesday	Christmas lights hunt	Jean	Aunt
Thursday	Christmas crafts	Albert	Grandma
Friday	Christmas movies	Gary	Grandpa
Saturday	Board games	Tess	Uncle

DECEMBER 22

SCENTED SECRETS

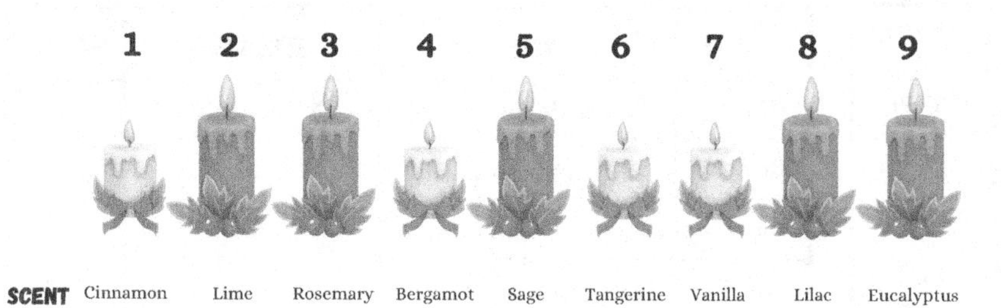

	1	2	3	4	5	6	7	8	9
SCENT	Cinnamon	Lime	Rosemary	Bergamot	Sage	Tangerine	Vanilla	Lilac	Eucalyptus

HIKE & SEEK

DAY	TRAIL	MILES	ANIMAL
Monday	Ember Valley	7	Elk
Tuesday	Avalanche Pass	3	Owl
Wednesday	Sapphire Summit	6	Marmot
Thursday	Thunderbolt Walk	5	Red fox
Friday	Frostbite Ridge	4	Mountain lion

DECEMBER 23

WE THE PEOPLE

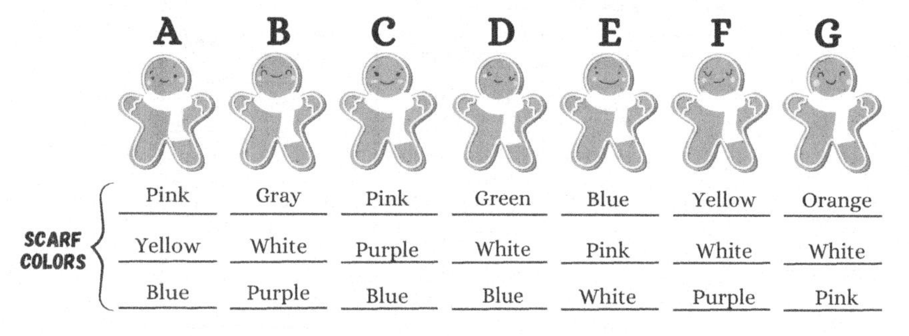

	A	B	C	D	E	F	G
SCARF COLORS	Pink	Gray	Pink	Green	Blue	Yellow	Orange
	Yellow	White	Purple	White	Pink	White	White
	Blue	Purple	Blue	Blue	White	Purple	Pink

SEASONAL STALLS

PRODUCT	SELLER	STAND	DECORATIONS
Artisanal soap	Olga	3	Garlands
Handcrafted ornaments	Steph	2	Fairy lights
Hand-poured candles	Nathan	1	Seasonal foliage
Hand-knit items	Elena	4	Snowflakes
Handmade pottery	Caleb	5	Tablecloths

DECEMBER 24

ASPEN'S LIFT LINE ENIGMA

	6	5	4	3	2	1
NAME	Arthur	Reuben	Elliot	Morgan	Isaac	Lacey
RIDES	16	13	17	12	14	15

BAKING UP HOLIDAY FUN

CHILD	SHAPE	COLOR	AGE	COOKIES
Devon	Star	White	6	17
Isabel	Snowman	Red	7	20
Luke	Candy cane	Yellow	9	21
Mike	Bell	Green	10	18
Yasmin	Snowflake	Blue	12	23

DECEMBER 25

THE GREAT CHRISTMAS CHALLENGE

RELATIVE	ACTIVITY	YEAR	MAIN COURSE	PHOTO ALBUMS	BOARD GAME
Aunt Mary	Decorate the tree	1988	Lobster thermidor	6	Scrabble
Aunt Patricia	Write letters to Santa	1983	Baked ham	2	Uno
Cousin Linda	Read Christmas stories	1987	Beef Wellington	3	Ticket to Ride
Grandma Evelyn	Play in the snow	1986	Stuffed pork loin	7	Monopoly
Grandpa Victor	Bake Christmas cookies	1989	Roast lamb	5	Dixit
Uncle Max	Scavenger hunt	1984	Seafood paella	9	Pictionary
Uncle Richard	Create Christmas crafts	1985	Prime rib roast	4	Risk

BONUS PUZZLES

Turn the page to step into a world of Christmas-themed puzzles suitable for all ages. Gather your family members and immerse yourselves in a festive atmosphere filled with logic and joy.

The answers await you on page 83 whenever you're ready to check your solutions.

Enjoy the holiday fun!

A SMALL REQUEST

Creating a book like this has been a labor of love, and as a small independent publisher, visibility is a constant challenge.

You can make a real difference by sharing your thoughts and experiences through a review on Amazon.

Your feedback not only supports us but also helps us continue to bring you more engaging and enjoyable books in the future.

Thank you for being a part of our community and for helping us thrive!

1. CHRISTMAS COOKIES

Emma spent the afternoon baking delicious Christmas cookies for her family.
After taking the last tray out of the oven, she counts the cookies: they are almost 100.
If only she had baked one more cookie, she could have divided them equally among herself, her mom, dad, and four siblings.

Emma's aunt Olivia shows up unexpectedly. Emma hides a cookie in her pocket and divides the remaining cookies equally among all the people present, including herself.

How many Christmas cookies did Emma bake?

2. A LONG LIST

Cathy wrote a long Christmas list for Santa. Here is the first page of her list.

Eight of the items in Cathy's list are things one can wear.
Ten of the items in the list have more than six letters.

What is the minimum number of items in Cathy's Christmas wish list?

3. CHRISTMAS SALES

My sister and I learned in school how to make beautiful Christmas decorations. Today we went around our neighborhood to sell the products of our hard work. I sold my decorations for $3 each, while my sister sold hers for $4 each.

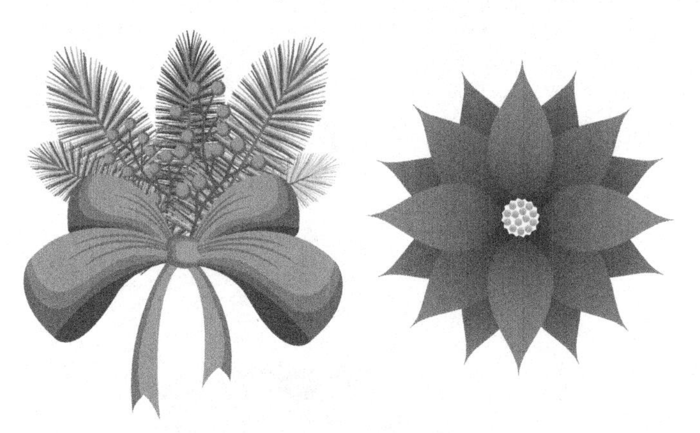

At the end of the day, we put together all the money we made: 17 dollars in total.

Who sold more Christmas decorations between me and my sister?

4. WINTER SPORTS

In a group of friends, 25 can ski and 21 can snowboard. The number of friends in the group is a multiple of 12.

How many people in that group can both ski and snowboard?

5. BUSY ELVES

Christmas is approaching and Santa's elves are wrapping presents as fast as they can.
Every elf has a pile of packages to wrap next to their desk.
At the moment, Alabaster Snowball has 3, Pepper Minstix has 5, and Sugarplum Mary has 7.

The head elf Humphrey walks in with a pile of 21 packages. He is not satisfied of the way they have been wrapped and wants them redone.
He distributes the 21 packages to Alabaster, Pepper, and Sugarplum so that now each elf has the same number of packages.

How many packages does Sugarplum Mary receive?

6. TIME FOR DIET

Santa really let himself go in the last few months and put on a lot of weight. His working clothes don't fit anymore, and he needs to order new ones. For this reason, he is now measuring his belly with a measuring tape.

We cannot reveal the results of the measurement for reasons of privacy, but we have a question for you. The tape measures 100 in. It has all the numbers from 1 to 100 on both sides, but on the back, they appear in reverse order.

Which number is written opposite 43?

7. CHRISTMAS CARDS

Over the years, Sarah has received lots of Christmas cards. Now she has a total of 250. Finally, she has bought some albums to keep them. Every album has 12 pages, and each page can contain 8 Christmas cards.

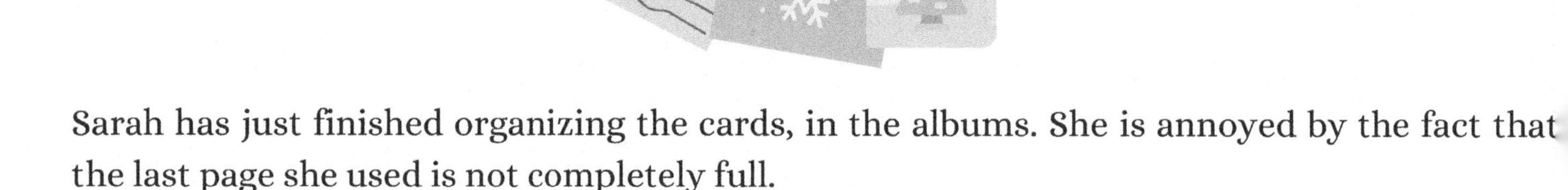

Sarah has just finished organizing the cards, in the albums. She is annoyed by the fact that the last page she used is not completely full.

Suddenly, the doorbell rings. It's the mailman, with some more Christmas cards. What a coincidence! They are just enough to complete that page.

How many Christmas cards did the mailman bring?

8. WHITE CHRISTMAS

This morning at 10, my brother and I watched the weather forecast on TV. They said that by 8 pm, 56 inches of snow would fall.

At 5 pm my brother went outside to measure the level of the snow and then told me, "If the weather forecast is correct, in the next three hours we will have an average of 5 inches of snow every 30 minutes".

How much snow fell today before 5 pm?

9. SANTA'S HELPERS

Santa calls an urgent meeting of all his staff. After just a few minutes, a good number of reindeer and elves have arrived, ready to listen to Santa's instructions.

Santa looks closely at his staff. He can count a total of 40 heads and 100 legs.

How many reindeer are present at the meeting?

10. SANTA'S TRIP

Santa is studying the map of a little town he must visit to deliver presents. The town has eight streets. At the intersections there are four houses.

Santa wants to start from one of the four houses in the map and go through every street once. (Apparently, all these eight streets are really pretty and worth seeing.) Of course, he will have to cross some of the intersections more than once.

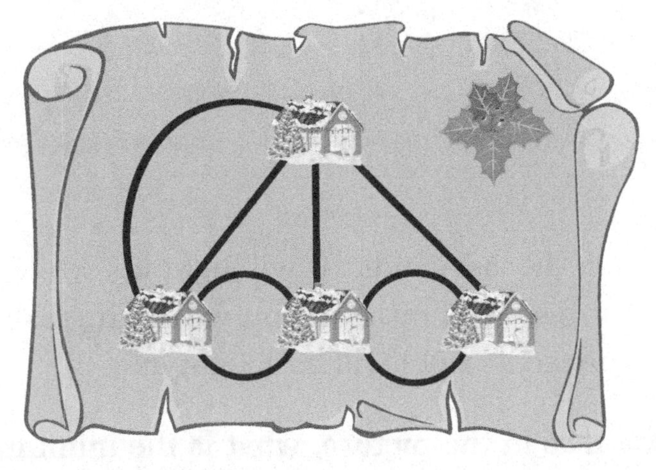

From which house will Santa start his visit of the town? At which house will he end it?

11. SWEET TRADING

My cousins and I always trade Christmas sweets. We have agreed on strict rules for our exchanges:

- One Christmas cupcake can be exchanged with two gingerbread cookies.
- One gingerbread cookie can be exchanged with two candy canes.
- One candy cane can be exchanged with two peppermint candies.

Before starting the exchanges, I had one cupcake, three candy canes, and two peppermint candies. At the end of the trading, I only had three items.

Which treats did I have after trading with my cousins?

12. CHRISTMAS CANDLES

My mom decorated the living room for Christmas with six red candles. However, she only li three of them, as you can see from the picture below.

These are very special candles: if you blow on an extinguished one, it lights up!
Moreover, whenever you blow on a candle, you also affect the candles immediately next to it.

For instance, if you blow on the last candle, it will light up, while the one to its left will go out. If you blow on the third candle, instead, you will extinguish it, but at the same time the one to its left and the one to its right will both light up.

Starting from the situation as in the picture, what is the minimum number of times you must blow on the candles before they are all extinguished?

13. A SACK OF CANDIES

Terrell, dressed as Santa, meets up with his friend Jacob, who is dressed as an elf. Terrell has a huge sack full of candies, which the two friends start sharing with the following procedure: Jacob takes one, then Terrell takes two, then Jacob takes three, and so on.

They go on in this way until, after Terrell's turn, there are no more candies left. At that point, they count their candies and find that Terrell has 10 more than Jacob.

How many candies were in the sack?

14. MATH DECORATIONS

Lucy stares at her Christmas tree in front of the fireplace. It's an amazing atmosphere, but to make it really perfect she wants to give a mathematical touch.

She starts decorating Christmas ornaments following a precise logic pattern. Here are the first five she produced.

At this point she stops. She thinks she might have to change her pattern.

Can you understand what is bothering Lucy?

15. COUNTDOWN

It's December, and Eric wants to know exactly how many days are missing before the beginning of the new year. On his calendar, he writes the number 1 on today's date, the number 2 on tomorrow's date, and so on.

Eric writes the last number inside the box corresponding to the first day of January.
At that point, he realizes that the number corresponding to today's date is equal to the total number of digits he wrote.

What is the date today?

16. FIREWORKS

The new year has just begun, and Ryan is at the window watching the fireworks.
In the meantime, his parents, aunts, and uncles are having a New Year's toast. Everyone clinks glasses with everyone else, and Ryan hears 15 clinks.

Ryan's mom comes to see the fireworks next to Ryan and misses the second toast.

How many clinks does Ryan hear this time?

17. GRANDMA'S RIDE

Grandma came to fetch the Christmas tree from my house to bring it to her place, where the entire family will spend the Christmas holidays.

Despite my father's protests, she loaded the tree over her car, saying that she will be extremely careful. She promised to drive the 60 miles that separate the two houses at an average speed of 10 miles per hour.

After 2 hours, grandma realizes that she has been going quite fast: she has already covered 30 miles.

What average speed must she keep during the rest of the trip to fulfil her promise?

18. MYSTERIOUS GIFTS

Daniel is a puzzle lover, and therefore Santa left him a little logic test.

Under the tree there are three packages. Each of them has a label to help Daniel understand where his present is. Santa also left a note warning that only one of the three labels tells the truth.

Which of the three packages should Daniel open to find his present?

19. HOLIDAY BREAD

Henry baked three loaves of mouth-watering sweet bread with sprinkles.

Henry's mom cuts each loaf into equal slices making a total of 15 vertical cuts. She then divides the slices equally among her children: Henry, Alexander, Mia, Evelyn, and Isabella.

How many slices of sweet bread will each child receive?

20. ENORMOUS PRESENTS

Santa left at my grandparents' house two huge presents for me and my sister. The combined height of the two packages is 22 feet! My dad has secured them on top of our car to bring them home.

Dad is worried since on the way home we must pass under a 40 ft bridge. Dad only managed to measure the height of the car with one single package on top. He tried with each of the packages and measured 25 ft and 27 ft, respectively.

Will the car manage to pass under the bridge with both packages on top?

21. HOLIDAY BAKING

As a joke, Cindy mixed up the labels of the six containers on the kitchen shelves. Now there is no container having the correct label.

Cindy's mom wants to start baking Christmas cookies. She asks Cindy where the flour is. Cindy replies, "I will only tell you that the salt is right under the jar with cocoa. Moreover, the sugar is next to the container with coffee, to the right".

Which container should Cindy's mom take from the shelf to find the flour?

22. ONE MORE REINDEER

Santa is trying to arrange geometrically all his reindeer. He tells them to form lines of three, but in that way, Rudolph is left out. They try with lines of four, but Rudolph is still left out.

Santa, exasperated, makes one last attempt, asking the reindeer to form lines of five. Once again, Rudolph doesn't find a spot and is the only one left out.

Knowing that there are less than 100 reindeer working for Santa, can you tell their exact number (including Rudolph, of course)?

23. SNOWMEN

Walter is outside playing in the snow with his sister Alice. Their favorite game is making huge snowmen.

It takes Alice 12 minutes to build a snowman. Walter is a bit slower, and he needs 24 minutes.

On the last snowman, they worked together. How long did it take them to build it?

24. GEOMETRY AND TREES

Elizabeth loves geometry almost as much as she loves Christmas.
Today, she decided to decorate her house with Christmas trees that follow a precise geometric pattern.

Elizabeth cannot resist the temptation to use her mathematical skills and calculate the area of the trees.
She has determined with certainty that the area of the first tree is 15 square feet and the area of the second is 23 square feet.

What is the area of the third tree?

25. HOLIDAY WISHES

The Johnsons send their 5 children to the Williams, who live at the end of the same street, to wish them happy holidays. The Williams also send their 5 children to the Johnsons for the same reason. The first child of each family leaves their house at 11 am. Every other child leaves one minute after the previous one.

It takes 3.5 minutes to go from one house to the other. Whenever two children cross path with each other they both say, "Merry Christmas."

How many times did the ten children say "Merry Christmas" during their walk from house to house?

26. A PERFECT TREE

Charlotte and Sophie know exactly which tree they want to buy for Christmas. It's truly majestic, and they are looking forward to having it in their garden and decorate it.

Unfortunately, the tree is expensive. "I would only need two more dollars to buy it," says Charlotte. Sophie searches her pockets and finds some money "I have very little here. I would need 47 more dollars to buy it". The two girls put their money together, but still don't have enough to buy the tree.

Knowing that the price of the tree is a whole number of dollars, can you determine how much it costs?

27. FUNDRAISING

Julie wants to sell a basket full of beautiful Christmas decorations to her family members as part of a fundraising campaign.

First, Julie goes to her aunt Mildred, who buys half of the decorations. Then it's the turn of uncle Dorian, who buys 5 pieces. Julie's grandma buys half of what is left, and then Julie's sister Amanda buys the last two remaining decorations.

How many decorations did Julie have in her basket when she went to aunt Mildred?

28. ICE SKATING

Lillian and Zoe start ice skating at the same time from the same point.

Lillian is very skilled, and it takes her 5 minutes to go around the ice-skating rink once. Zoe is a bit slower and covers the same distance in 7 minutes.

After a while, Lillian passes Zoe. After some more time she passes her again, and then again. When Lillian is passing Zoe for the fourth time, they decide that they are tired and should go drink a hot chocolate.

For how long were Lillian and Zoe ice skating?

29. SINGING IN THE SNOW

Berty, Scout, and Igloo are three very talented penguins. Their specialty is singing Christmas carols.

The three penguins' repertoire consists of five different carols. Two of these five songs, however, are actually the first and second part of a single carol. Therefore, they should be performed one after the other in that specific order.

In how many different orders can Berty, Scout, and Igloo perform their five songs?

30. CHRISTMAS FRILLS

Jennifer is using a scale to weigh her favorite Christmas ornaments. Here are three situations in which the scale is perfectly balanced.

How many balls are needed to match the weight of the star?

31. PAIR THE ORNAMENTS

In this game, your objective is to link every pair of identical ornaments using a path comprising exclusively of horizontal or vertical segments. It's essential that no two paths intersect. Additionally, every vacant box within the grid must be traversed by a line.

Example

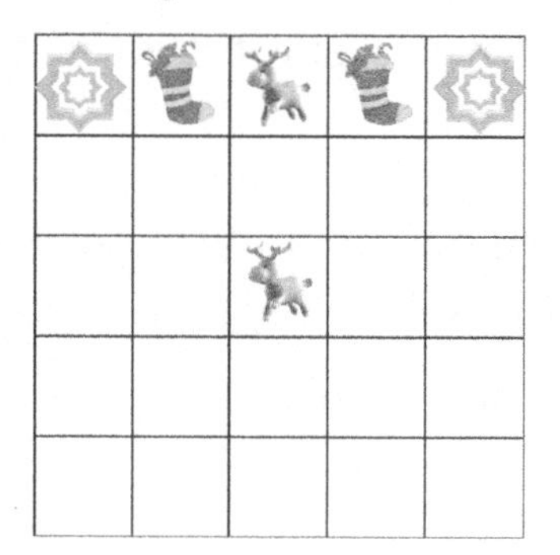

Solution

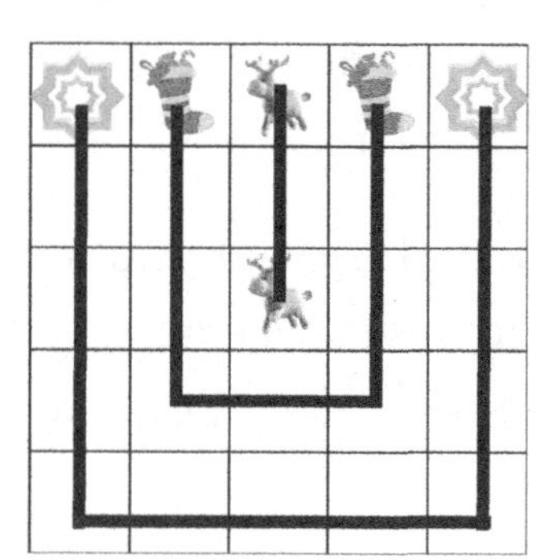

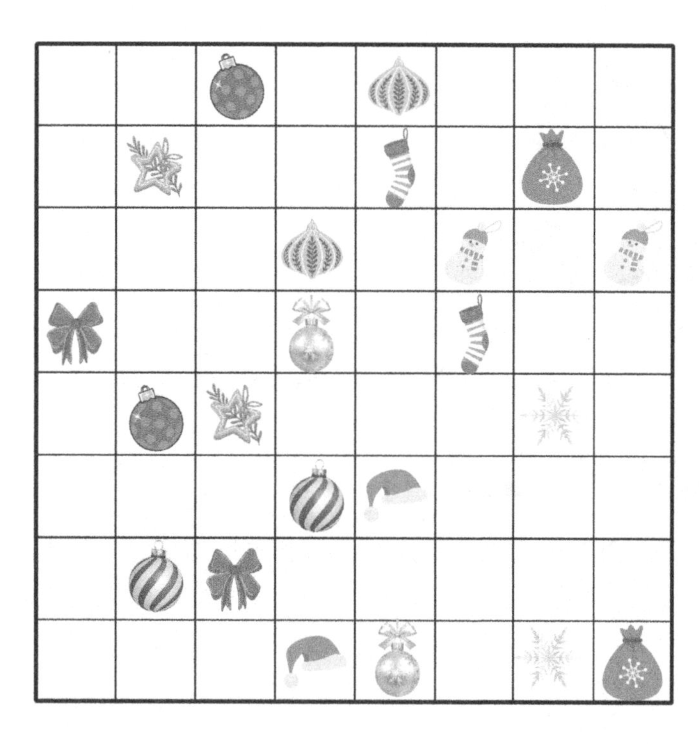

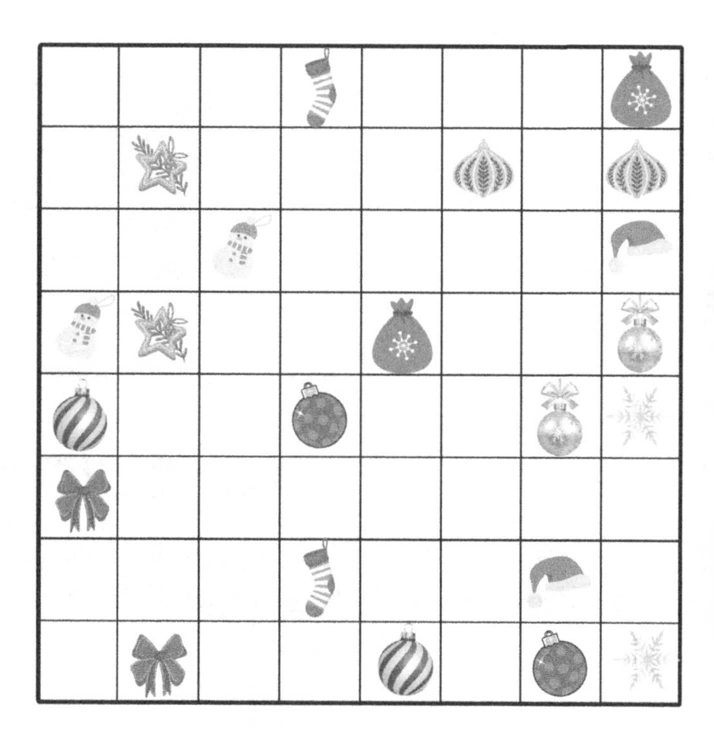

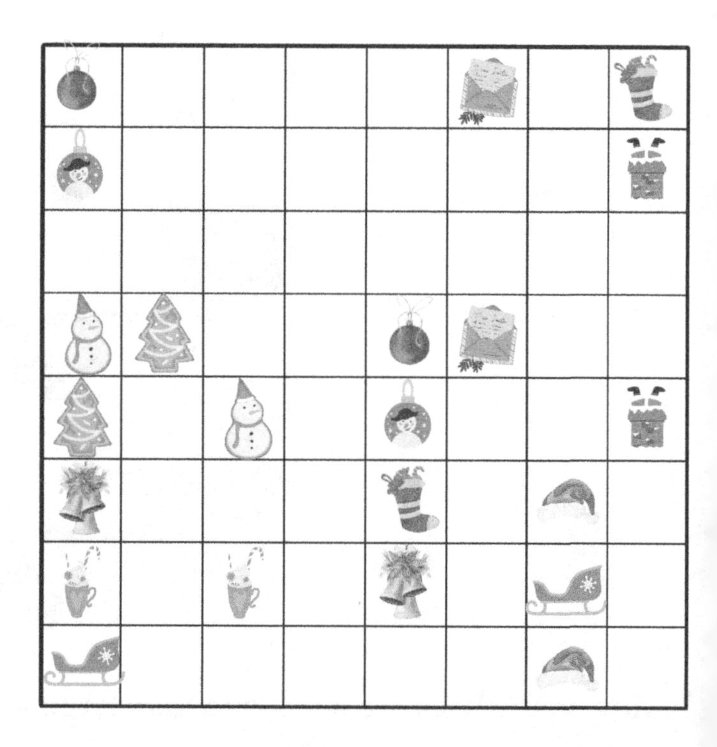

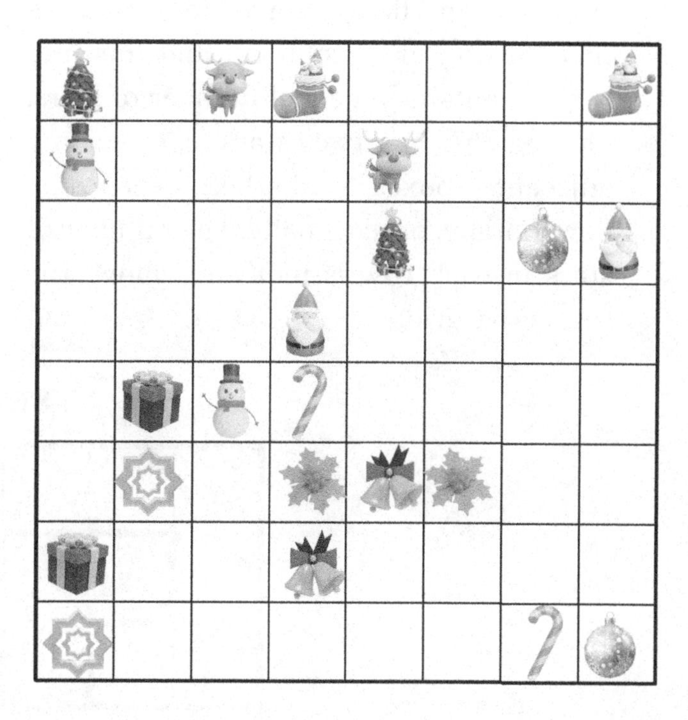

32. TWINKLE TWINKLE LITTLE STAR

In this puzzle, your objective is to place stars within the grid. You must ensure that every row, column, and distinct region contains an equal number of stars. The catch? No two stars can occupy adjacent boxes, whether vertically, horizontally, or diagonally. The number of stars in each row is indicated above the respective grid.

Example

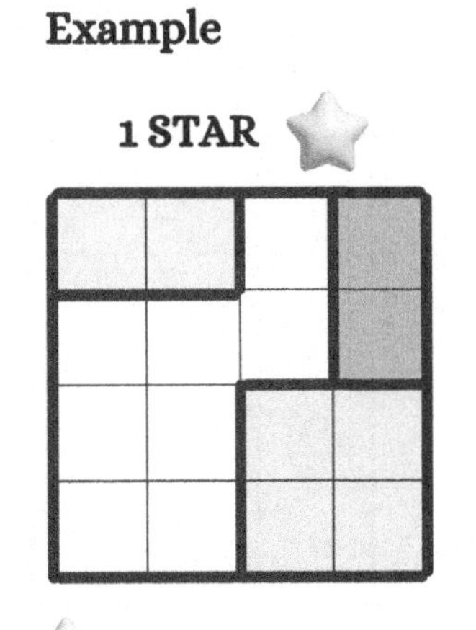

Solution

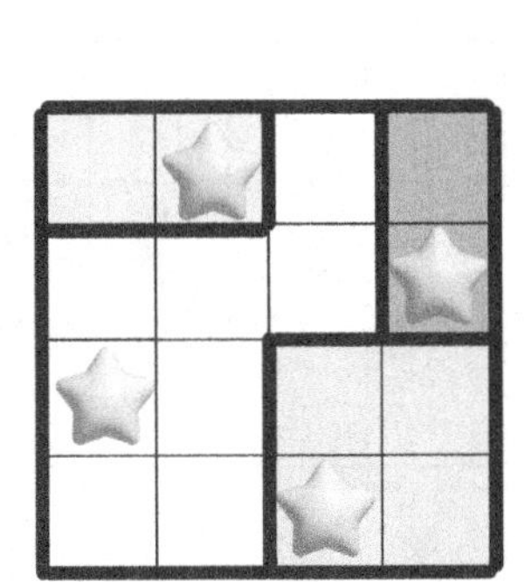

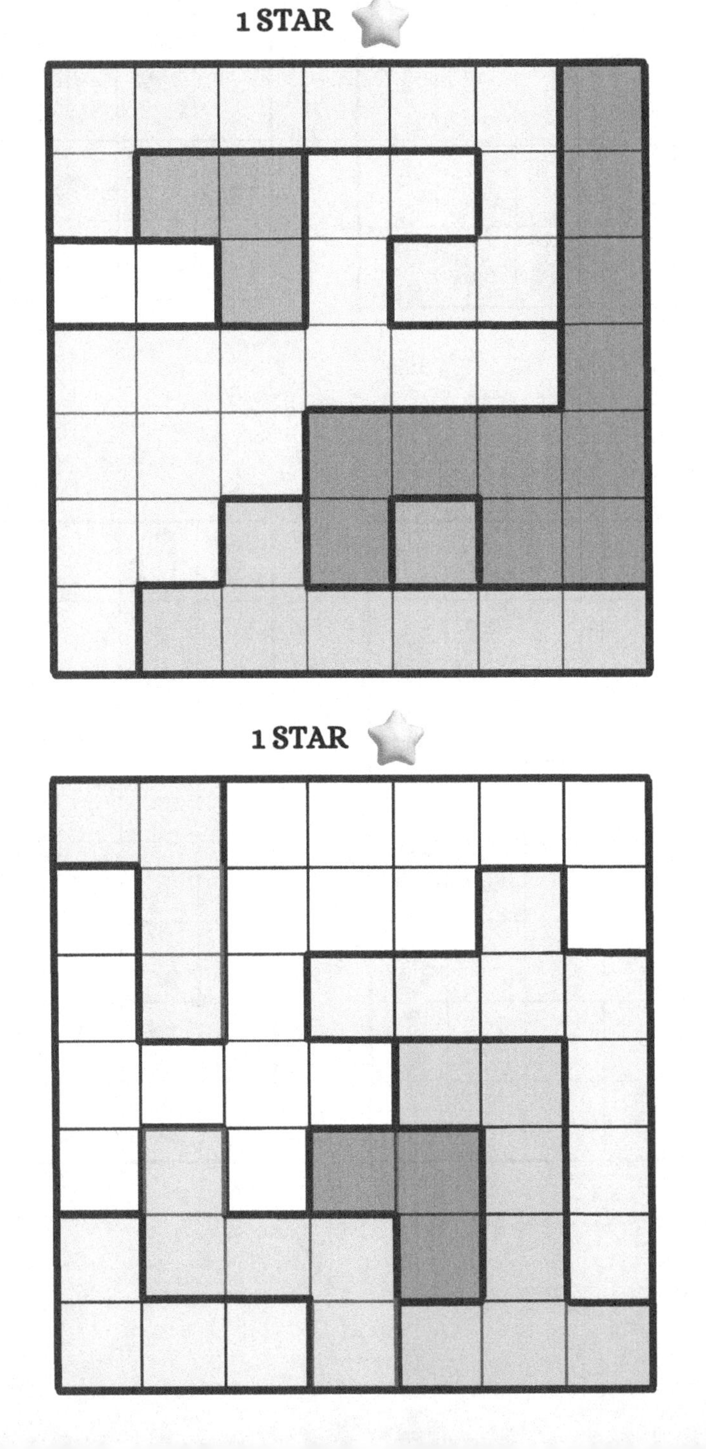

2 STARS

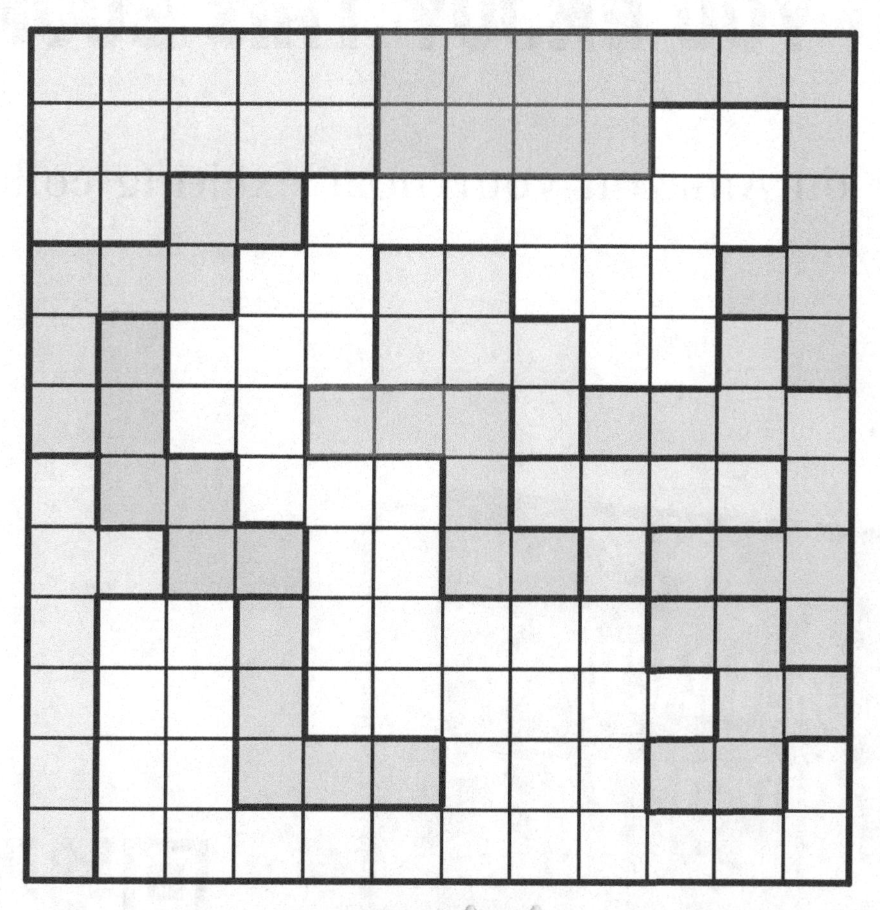

2 STARS

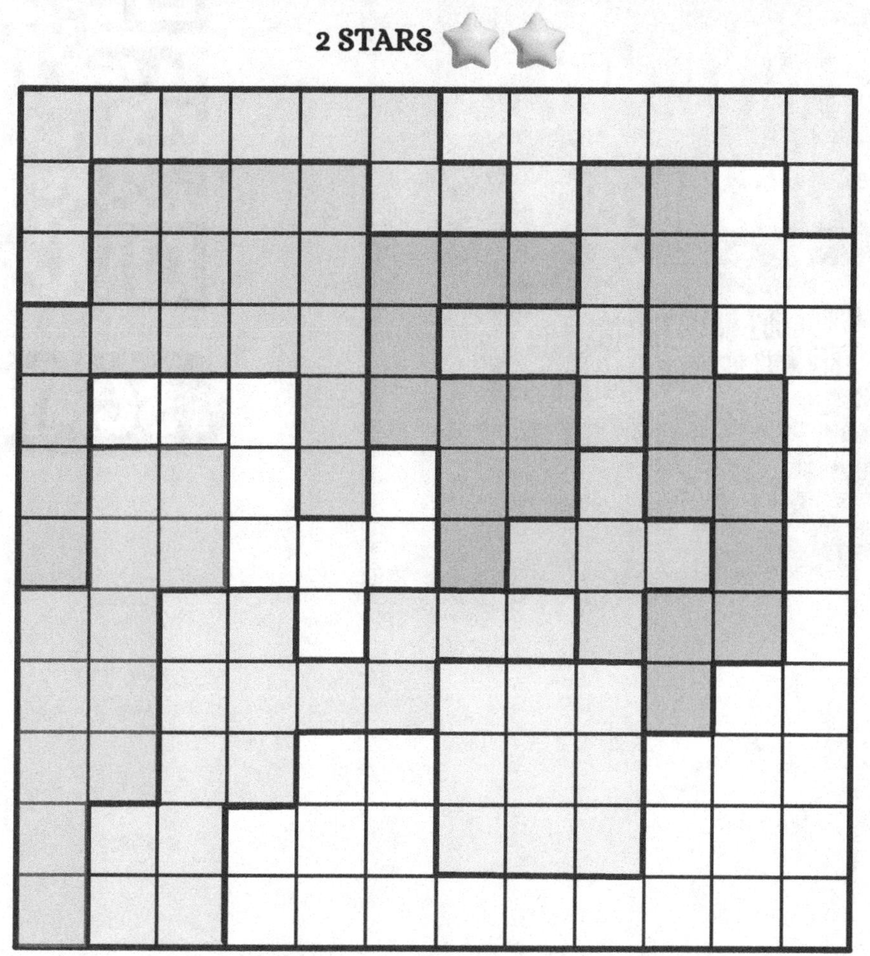

DID YOU ENJOY THIS BOOK?

Buy now on Amazon your next exciting collection of logic puzzles.

SOLUTIONS

1. CHRISTMAS COOKIES

Emma baked 97 Christmas cookies. If she had baked one more cookie, she would have 98. This number is divisible by 7 (Emma, her parents, and her four siblings). After hiding one cookie in her pocket, she has 96, which is divisible by 8.

2. A LONG LIST

The minimum number of items in Cathy's Christmas wish list is 12: the 5 elements in the first page plus 7 more. These 7 additional items are all words with more than 6 letters. Four of them also belong to the category "things one can wear."

3. CHRISTMAS SALES

I sold 3 decorations, earning $9. My sister sold 2, earning $8.

4. WINTER SPORTS

There are 10 people who can both ski and snowboard. First note that the minimum number of people in the group is 25 (in case everyone who can snowboard can also ski). The maximum number of people is 46 (in the case where nobody can both ski and snowboard). The number of people in the group is 36, since this is the only multiple of 12 between 25 and 46. Adding the number of people who can snowboard to the number of people who can ski gives 46. Since there are only 36 people, 10 people have been counted twice.

5. BUSY ELVES

Sugarplum Mary receives 5 packages.

6. TIME FOR DIET

The number written opposite 43 is 57. In fact, the sum of two numbers on opposite sides of the tape is always 100.

7. CHRISTMAS CARDS

The mailman brought 6 Christmas cards.

8. WHITE CHRISTMAS

At 5 pm, we had 26 inches of snow.

9. SANTA'S HELPERS

At the meeting, there are 10 reindeer.

10. SANTA'S TRIP

Consider the house on the bottom right and the one immediately to its left. Santa can start from either of these two and end on the other.

11. SWEET TRADING

After the exchanges I had one Christmas cupcake and two gingerbread cookies.

12. CHRISTMAS CANDLES

I must blow on at least three candles. Here is a possible solution.

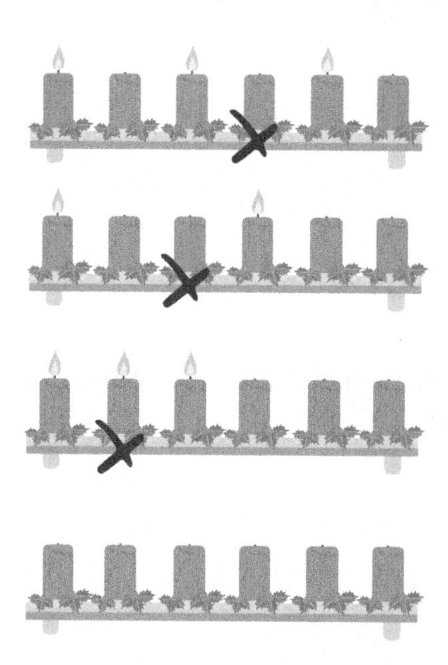

13. A SACK OF CANDIES

Every turn, Terrell picks one candy more than Jacob. Therefore, there were 10 turns in total. In the sack there were 210 candies. In fact, 210=1+2+3+4+...+20.

14. MATH DECORATIONS

On every ornament, Lucy writes a number corresponding to the number of letters written on the previous ornament. If she were to continue in this way, she would have to keep writing "four" on every subsequent ornament.

15. COUNTDOWN

Today is December 19. To get to January first, Eric has to write all the numbers from 1 to 14, which requires a total of 19 digits.

16. FIREWORKS

There were 6 people for the first toast. After Ryan's mom leaves, there are 5 people. The number of clinks is now 10.

17. GRANDMA'S RIDE

Grandma should go at 7.5 miles per hour. In that way, she will cover the remaining 30 miles in 4 hours.

18. MYSTERIOUS GIFTS

The present is in the blue package.

19. HOLIDAY BREAD

There are 18 slices in total. Each child receives 3 slices.

20. ENORMOUS PRESENTS

The total height of the car with the two packages on top is (22+25+27)/2=37 ft. In fact, in the sum 22+25+27, every element (the car and the two packages)is counted twice. The car will therefore manage to pass under the bridge.

21. HOLIDAY BAKING

The flour is in the jar labeled "Sugar".

22. ONE MORE REINDEER

Excluding Rudolph, the number of reindeer is divisible by 3,4, and 5. The only such number between 1 and 100 is 60. Adding Rudolph, we obtain a total of 61 reindeer.

23. SNOWMEN

In 1 minute, Alice makes 1/12 of a snowman, while Walter 1/24 of a snowman. When working together, they make 1/12 + 1/24 = 1/8 of a snowman every minute. Therefore, it takes Alice and Walter 8 minutes to make a snowman when they work together.

24. GEOMETRY AND TREES

The third tree is equal to the second one plus the difference between the first two. Therefore, its area is 23+(23-15) = 31 square feet.

25. HOLIDAY WISHES

The first Johnson child meets all but one of the Williams children. The next three Johnsons meet all five the Williams children. The last Johnson meets four of the Williams children (missing the one who left first). In total, there were 4+5+5+5+4=23 encounters. Therefore, "Merry Christmas" was said 46 times.

26. A PERFECT TREE

Sophie must have less than two dollars, otherwise putting together their money, the two girls would be able to buy the tree. Since the tree costs a whole number of dollars, Sophie has one dollar and the tree costs $48.

27. FUNDRAISING

Reasoning backwards, one deduce that Julie had 18 decorations when she went to Aunt Mildred.

28. ICE SKATING

Lillian and Zoe were ice skating for 70 minutes. Every 35 minutes Lillian goes around the ice-skating rink 7 times, while Zoe only 5. Therefore, Lillian passes Zoe for the fourth time after 70 minutes.

29. SINGING IN THE SNOW

Since two songs must be performed one after the other in that specific order, the problem is equivalent to the three penguins having only four songs in their repertoire. Berty, Scout, and Igloo can perform their songs in 24 different orders.

30. CHRISTMAS FRILLS

Six balls are needed to match the weight of the star.

31. PAIR THE ORNAMENTS

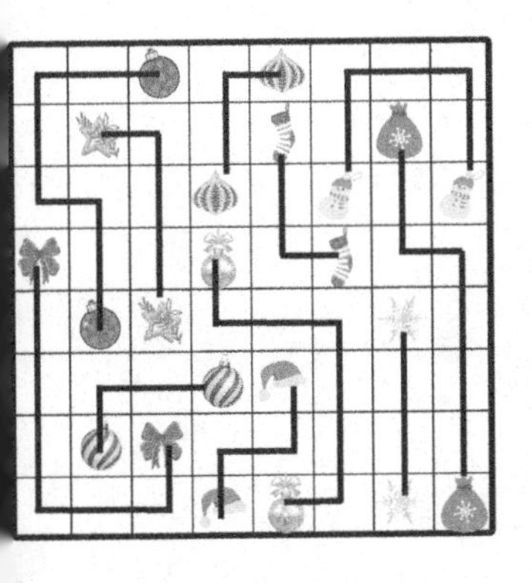
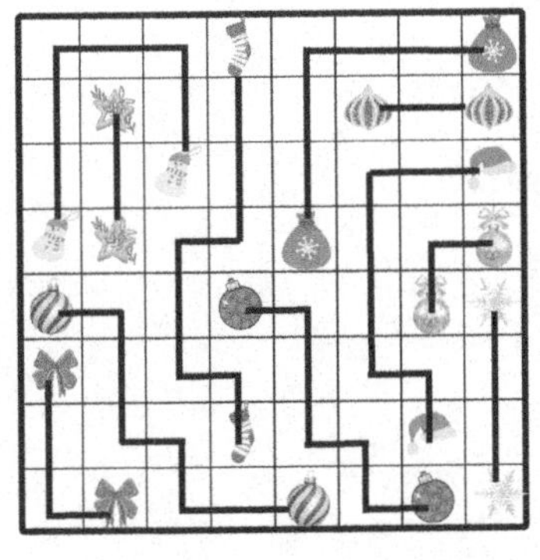
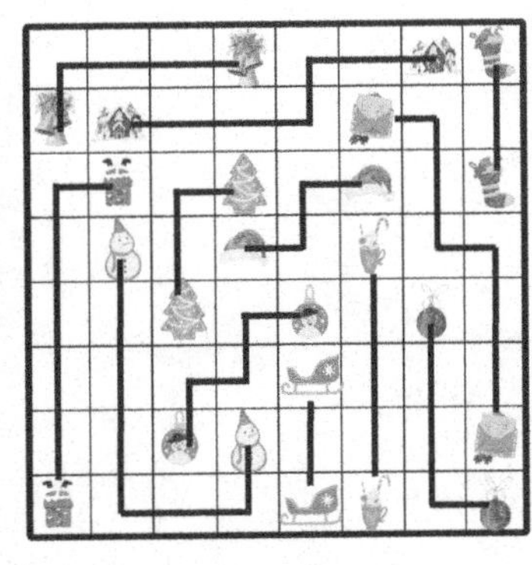
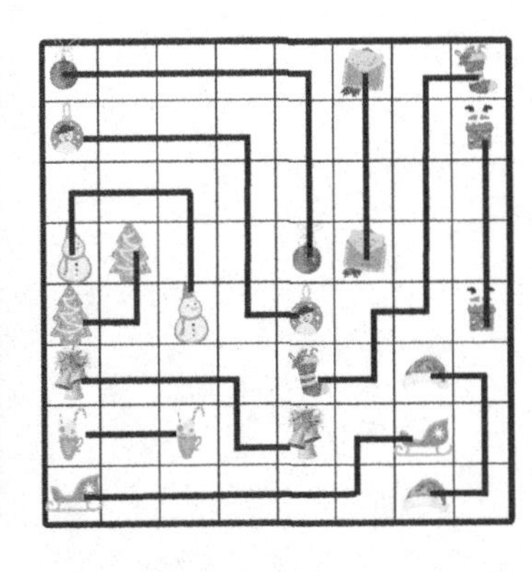

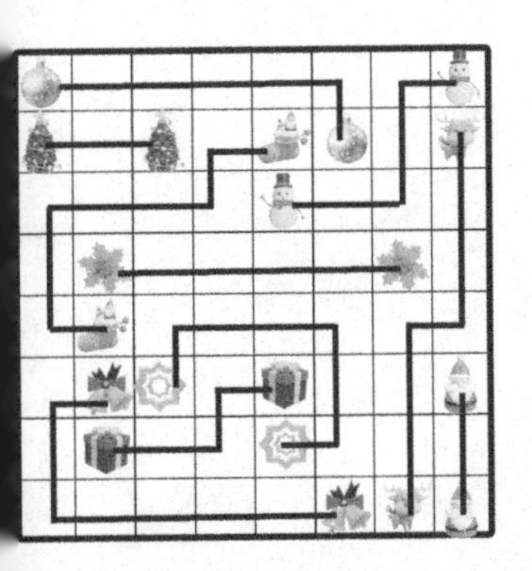
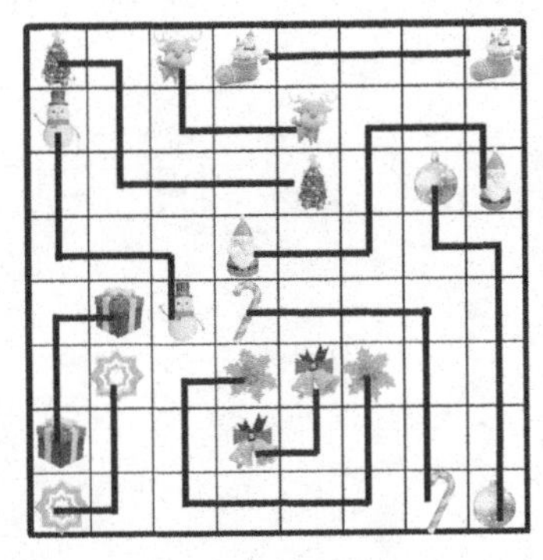
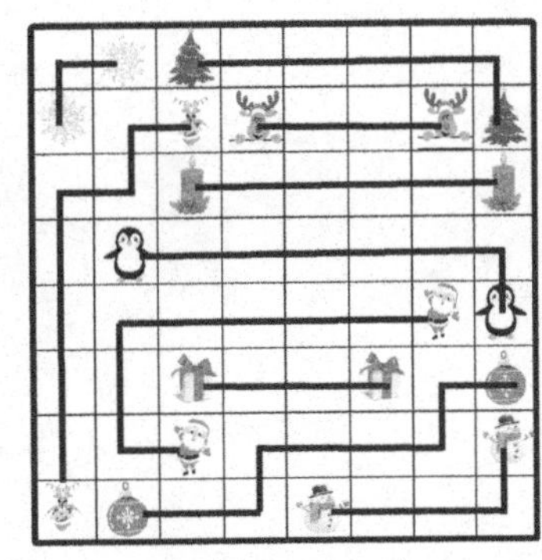
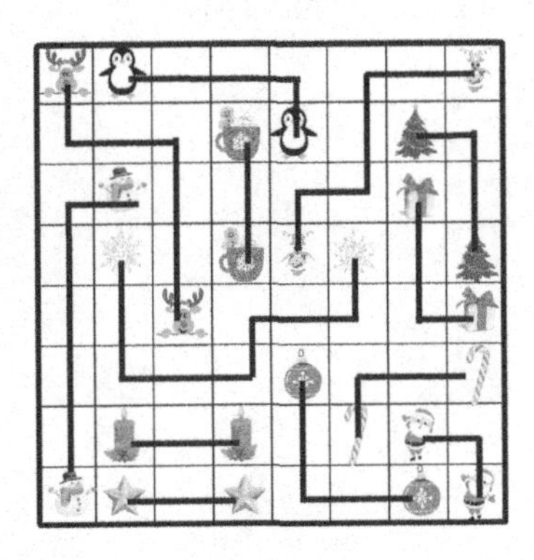

32. TWINKLE TWINKLE LITTLE STAR

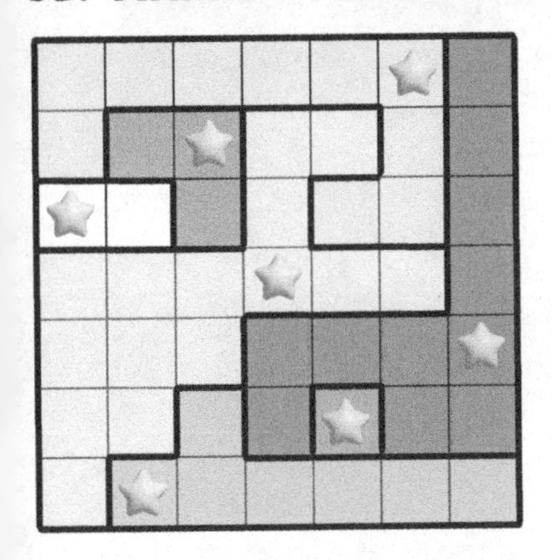
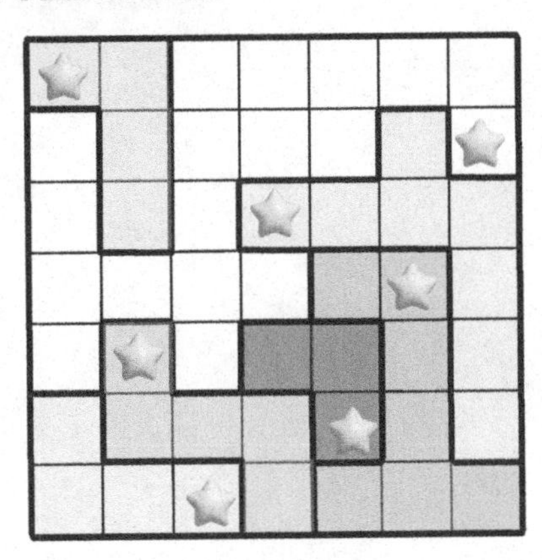

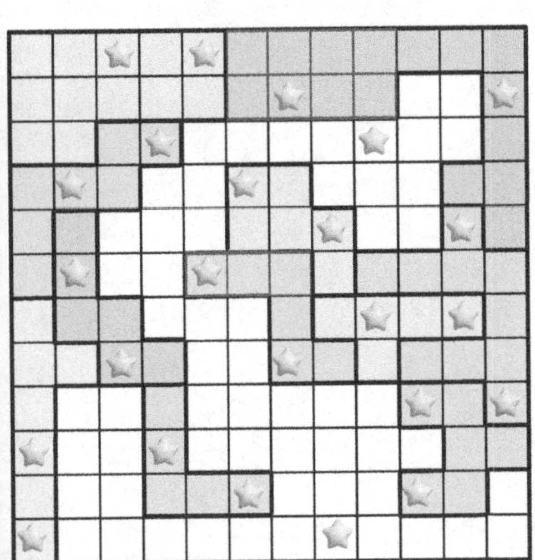

Made in the USA
Las Vegas, NV
24 November 2023